T0306009

Nanotubes, Nanowires, Nanobelts and Nanocoils— Promise, Expectations and Status

MATERIALS RESEARCH SOCIETY
SYMPOSIUM PROCEEDINGS VOLUME 1142

Nanotubes, Nanowires, Nanobelts and Nanocoils— Promise, Expectations and Status

Symposium held December 1–4, 2008, Boston, Massachusetts, U.S.A.

EDITORS:

Prabhakar R. Bandaru

University of California, San Diego
La Jolla, California, U.S.A.

Sonia Grego

RTI International
Research Triangle Park, North Carolina, U.S.A.

Ian Kinloch

University of Manchester
Manchester, United Kingdom

Materials Research Society
Warrendale, Pennsylvania

CAMBRIDGE
UNIVERSITY PRESS

University Printing House, Cambridge CB2 8BS, United Kingdom

One Liberty Plaza, 20th Floor, New York, NY 10006, USA

477 Williamstown Road, Port Melbourne, VIC 3207, Australia

314-321, 3rd Floor, Plot 3, Splendor Forum, Jasola District Centre, New Delhi - 110025, India

79 Anson Road, #06-04/06, Singapore 079906

Cambridge University Press is part of the University of Cambridge.

It furthers the University's mission by disseminating knowledge in the pursuit of
education, learning and research at the highest international levels of excellence.

www.cambridge.org
Information on this title: www.cambridge.org/9781605111148

Materials Research Society
506 Keystone Drive, Warrendale, PA 15086
http://www.mrs.org

First published 2009
First paperback edition 2012

Single article reprints from this publication are available through
University Microfilms Inc., 300 North Zeeb Road, Ann Arbor, MI 48106

CODEN: MRSPDH

A catalogue record for this publication is available from the British Library

ISBN 978-1-605-11114-8 Hardback
ISBN 978-1-107-40837-1 Paperback

CONTENTS

*NANOWIRES: SYNTHESIS AND
CHARACTERIZATION*

CARBON NANOSTRUCTURE PROCESSING

PREFACE

   Lower dimensional structures, such as nanotubes and nanowires, have opened up new frontiers in materials sciences. Recent advances in their synthesis, processing and characterization indicate that some of their enormous potential is slowly being realized. Symposium JJ, "Nanotubes, Nanowires, Nanobelts and Nanocoils—Promise, Expectations and Status," held December 1–4 at the 2008 MRS Fall Meeting in Boston, Massachusetts, aimed to critically review the progress that had been made in the science and applications in both elemental (*e.g.,* C and Si) and compound (*e.g.,* ZnO, InP and GaAs) nanotubes, wires and layers. The symposium emphasized how far the initial expectations have come to fruition and the work necessary in the future to realize their promise. These proceedings highlight some of the papers presented during the conference covering these themes.

Prabhakar R. Bandaru
Sonia Grego
Ian Kinloch

March 2009

MATERIALS RESEARCH SOCIETY SYMPOSIUM PROCEEDINGS

MATERIALS RESEARCH SOCIETY SYMPOSIUM PROCEEDINGS

Prior Materials Research Society Symposium Proceedings available by contacting Materials Research Society

Carbon Nanotube Growth Mechanisms

Mater. Res. Soc. Symp. Proc. Vol. 1142 © 2009 Materials Research Society   1142-JJ02-02

In Situ Observation of Nucleation and Growth of Carbon Nanotubes from Iron Carbide Nanoparticles

Hideto Yoshida[1], Seiji Takeda[1], Tetsuya Uchiyama[1], Hideo Kohno[1], and Yoshikazu Homma[2]
[1]Department of Physics, Graduate School of Science, Osaka University, 1-1 Machikaneyama, Toyonaka, Osaka 560-0043, Japan
[2]Department of Physics, Tokyo University of Science, Shinjuku, Tokyo, 162-8601, Japan

ABSTRACT

Nucleation and growth processes of carbon nanotubes (CNTs) in iron catalyzed chemical vapor deposition (CVD) have been observed by means of *in-situ* environmental transmission electron microscopy. Our atomic scale observations demonstrate that solid state iron carbide (Fe$_3$C) nanoparticles act as catalyst for the CVD growth of CNTs. Iron carbide nanoparticles are structurally fluctuated in CVD condition. Growth of CNTs can be simply explained by bulk diffusion of carbon atoms since nanoparticles are carbide.

INTRODUCTION

Carbon Nanotubes (CNTs) are most important and promising materials for future nanotechnology [1,2]. Recently, the growth methods of CNTs, in particular catalytic chemical vapor deposition (CVD), are developed and therefore we can obtain high purity, vertically aligned CNTs in large quantities [3]. In the CVD growth of CNTs, it is inferred that metal nanoparticles act as catalyst; however the details of the role are still unknown. In order to elucidate the CVD growth mechanism of CNTs including the role of nanoparticle catalysts (NPCs), *in-situ* observation of the CNT growth is one of the most promising methods [4-9]. In this study, we observe the nucleation and growth of multi-walled CNTs (MWNTs) from NPCs of fluctuating crystalline Fe carbide by atomic-scale *in-situ* environmental transmission electron microscopy (ETEM) [10].

EXPERIMENT

Iron was deposited as a catalyst on silicon substrates with thin SiO$_2$ surface layer by vacuum evaporation. The samples were set in a newly designed ETEM (FEI Tecnai F20 equipped with an environmental-cell) operated at 200 kV. The substrates were heated to 600 °C in a vacuum, and then a mixture gas of C$_2$H$_2$:H$_2$ = 1:1 was introduced into the ETEM. The pressure of the gas and the temperature of the substrates in the CVD condition in the ETEM were 10 Pa and 600 °C, respectively. The growth of CNTs was recorded at a rate of 1 frame per 0.35 s using a CCD camera. All the images presented in this paper are extracted from *in-situ* ETEM movies in the CVD condition.

RESULTS AND DISCUSSION

Nucleation of MWNTs

We have observed the nucleation of a MWNT from a NPC in CVD condition (Fig. 1) [10]. Individual graphene layers can be observed with sufficient spatial resolution. Firstly, graphene layers are formed on an apparent facet of the NPC (Fig. 1a). The graphene layers gradually extend in plane, and bend along the facets of the NPC (Fig. 1b). Additional graphene layers nucleate between the NPC and the existing ones in succession. During this process, the NPC is gradually deformed and a characteristic protrusion appears and extends further (Fig. 1a and 1e). The protrusion suddenly shrinks and a MWNT is expelled from the NPC (Fig. 1f and 1g). The root-grown MWNT is encapsulated at the tip. In Fig. 1g, the NPC exhibits clear lattice image, and the corresponding Fourier transform (Figure 1h) shows characteristic diffractions which cannot be explained by neither pure iron α(bcc) nor γ(fcc) structures.

Growth of MWNTs from iron carbide NPCs

Figures 2 and 3 show the root growth processes of MWNTs viewed nearly end-on and nearly normal to the growth direction, respectively. As shown in Fig. 2, a MWNT grows from a NPC that keeps exhibiting the same lattice image for 1.4 s. The lattice spacing in Fig. 2 is about

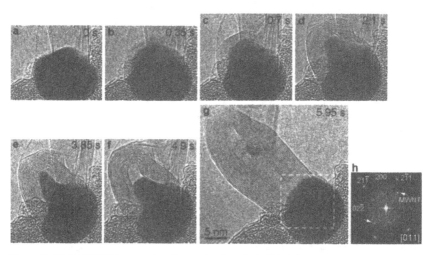

Figure 1. (a)-(g) ETEM images showing nucleation of a MWNT from a NPC on a substrate. Graphene layers nucleate on the NPC and the MWNT is expelled from the deformed NPC. The recording time is shown in images. (h) The Fourier transform of the region surrounded by the dotted line in (g). This corresponds to the [0$\bar{1}$1] zone-axis pattern from iron carbide (cementite, Fe_3C). The spots from the MWNT are marked by the arrows.

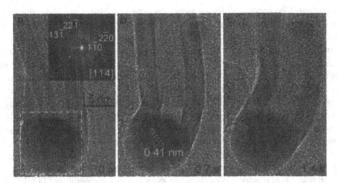

Figure 2. ETEM images showing a growing MWNT from an iron carbide NPC on a substrate viewed nearly end-on. The recording time is shown in images. Fourier transform of the dotted square region in (a) is also shown.

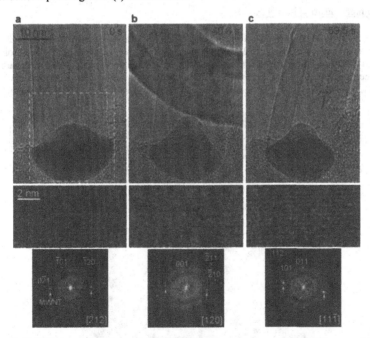

Figure 3. ETEM images showing a growing MWNT from an iron carbide NPC on a substrate viewed nearly normal to the growth direction. The recording time is shown in images. Enlarged images and Fourier transforms of the dotted square regions in images are shown below. The spots from the MWNT are marked by the arrows.

0.41 nm, which cannot be accounted for by assuming pure iron. Next, we show the growth of MWNT for a long time. Lattice image of the NPC in Fig. 3 is frequently smeared out, and another kind of lattice image reappears. All the lattice images that appear in the series of in Figs. 2 and 3 are consistently accounted for by the iron carbide, i.e. cementite, Fe_3C. The lattice image in Figure 1g can also be explained by the cementite structure. Based on the cementite structure, we can index the diffraction spots in Fourier transform of lattice images and determine the incident direction of the electron beam correspondingly. The incident beam directions are shown in Fig. 1h, Fig. 2a and Fig. 3. Measured interplanar spacing and angles in Fig. 1 to 3 agree well with those of Fe_3C within the accuracy of 1% and 2%, respectively. Fe_3C is the well-known metastable intermetalic compounds in the binary Fe-C system as a line phase. It is possible that the carbon concentration is slightly deviated from the stoichiometric composition since NPCs absorb and desorb carbon atoms dynamically during CNTs growth. These observations show that NPCs are iron carbide crystals and they fluctuate structurally during the growth of MWNTs. It is well known that metal nanoparticles fluctuate even at room temperature [11]. Much more surprisingly, the growth of MWNTs is much less interrupted by this structural fluctuation of carbide NPCs.

Growth mechanism of MWNTs

Structure and composition of the NPCs during CNT growth were controversial [12-13]. From the observations in Figs. 1 to 3, we can now clear the growth process of CNTs in iron catalyzed CVD condition. (1) NPCs are fluctuating crystalline particles. (2) NPCs are carbide such as cementite Fe_3C. (3) It is very likely that carbon atoms migrate through the bulk of NPCs since iron deposited on the substrate absorbs carbon and transforms to carbide NPCs in the

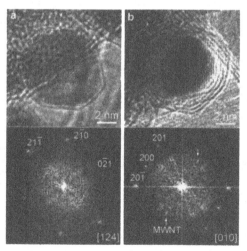

Figure 4. ETEM images of nanoparticles surrounded by (a) a graphene and (b) quadruple graphitic shell. The corresponding Fourier transforms are also shown. These nanoparticles can be identified as iron carbide (cementite, Fe_3C).

source gas of C_2H_2.. The third suggestion is supported by the observation that all the graphene cylinders of different diameter in a MWNT grow at the same growth rate. This means that carbon density is uniform at all the open ends of the graphene cylinders that are connected with the NPC. Hence, it is most likely that carbon atoms migrate toward the open ends through the NPC bulk rather than the surface of the NPC.

Figure 4 shows iron carbide nanoparticles surrounded by graphitic shells in the CVD condition. These nanoparticles are inactive for the growth of CNTs. The difference between active (as shown in Fig. 1 to 3) and inactive NPCs (in Fig. 4) still remains unclear. Our *in situ* observation as shown in Fig. 1, however, suggests that the deformation of NPCs is a crucial process for the nucleation of MWNTs. After the introduction of source gas, a graphitic layer nucleates at the surface of NPCs. When NPCs do not deform, the NPCs are covered with a graphitic shell entirely and CNTs never grow from such NPCs. The deformation of NPCs bends the graphitic layers and forms the cap of CNTs. This indicates that CNTs nucleate via a complex dynamic motion of carbon atoms and carbide NPCs, and that the yarmulke mechanism [14] is oversimplified.

CONCLUSIONS

Our atomic-scale *in situ* observation of CVD growth of CNTs has shown that structurally fluctuating iron carbide (Fe_3C) nanoparticles act as catalyst. In addition, we strongly suggest that carbon atoms migrate through the bulk of nanoparticle catalysts. These findings may bring general understanding of catalyzed CVD growth of CNTs at atomic scale.

ACKNOWLEDGMENTS

We thank Yusuke Tanemoto for preparing silicon substrates. This work was supported by CREST, JST.

REFERENCES

1. S. Iijima, *Nature* **354**, 354 (1991).
2. S. Iijima and T. Ichihashi, *Nature* **363**, 603 (1993).
3. K. Hata, D. N. Futaba, K. Mizuno, T. Namai, M. Yumura, and S. Iijima, *Science* **306**, 1362 (2004).
4. S. Helveg, C. López-Cartes, J. Sehested, P. L. Hansen, B. S. Clausen, J. R. Rostrup-Nielsen, F. Abild-Pedersen, and J. K. Nørskov, *Nature* **427**, 426 (2004).
5. R. Sharma and Z. Iqbal, *Appl. Phys. Lett.* **84**, 990 (2004).
6. H. Yoshida and S. Takeda, *Phys. Rev. B* **72**, 195428 (2005).
7. M. Lin, J. P. Y. Tan, C. Boothroyd, K. P. Loh, E. S. Tok, and Y.-L. Foo, *Nano Lett.* **6**, 449 (2006).
8. S. Hofmann, R. Sharma, C. Ducati, G. Du, C. Mattevi, C. Cepek, M. Cantoro, S. Pisana, A. Parvez, F. Cervantes-Sodi, A. C. Ferrari, R. Dunin-Borkowski, S. Lizzit, L. Petaccia, A. Goldoni, and J. Robertson, *Nano Lett.*, **7**, 602 (2007).
9. H. Yoshida, T. Uchiyama, and S. Takeda, *Jpn. J. Appl. Phys.*, **46**, L917 (2007).

10. H. Yoshida, S. Takeda, T. Uchiyama, H. Kohno, and Y. Homma, *Nano Lett.*, **8**, 2082 (2008).
11. S. Iijima and T. Ichihashi, *Phys. Rev. Lett.* **56**, 616 (1986).
12. C. Emmenegger, J.-M. Bonard, P. Mauron, P. Sudan, A. Lepora, B. Grobety, A. Züttel, and L. Schlapbach, *Carbon*, **41**, 539 (2003).
13. Y. H. Jung, B. Wei, R. Vajtai, P. M. Ajayan, Y. Homma, K. Prabhakaran, and T. Ogino, *Nano Lett.* **3**, 561 (2003).
14. H. Dai, A. G. Rinzler, P. Nikolaev, A. Thess, D. T. Colbert, and R. E. Smalley, *Chem. Phys. Lett.* **260**, 471 (1996).

Solution Based Processing
and Electrochemistry

Mater. Res. Soc. Symp. Proc. Vol. 1142 © 2009 Materials Research Society   1142-JJ04-03

Towards Novel Entangled Carbon Nanotube Composite Electrodes

Sherrell, P[1], Chen, J[1], Wallace, G G[1] Minett, AI[1]

[1]Intelligent Polymer Research Institute, ARC Centre of Excellence for
Electromaterials Science, University of Wollongong, Wollongong, NSW, Australia, 2522
aminett@uow.edu.au

ABSTRACT

The commercialization of carbon nanotube electrodes is impeded by the lack of bulk
processing techniques. One approach to overcome this impediment is the growth of macroscopic
CNT composite architectures which do not require any extra processing. Unfortunately the
fundamental growth mechanisms of these carbon composites is not currently understood. To
probe this mechanism a systematic examination of the effect of certain growth parameters was
undertaken. Within this paper we present the promising preliminary findings of this study
revealing extremely complex relationships between variables during growth. We also present the
performance of the produced architectures as capacitor electrodes and the further improvement
of these electrodes by doping with metallic nanoparticles.

INTRODUCTION

Carbon nanotubes (CNTs) have been the subject of intense research in terms of novel
electrode materials, primarily due to the high conductivities and surface area's attainable with
CNT structures.[1-5] A key limitation for many CNT applications is laborious post-processing
regimes required to generate a suitable structure. Recent developments in chemical vapor
deposition (CVD) of CNTs has lead to numerous 3D CNT architectures that can be utilized 'as
grown'.[6-9] Due to the catalytic requirements of many electrochemical devices, specifically
lithium ion (Li-ion) batteries and fuel cells (FCs), the doping of CNTs with metallic and metal-
oxide nanoparticles (NPs) is crucial to device applications of CNT electrodes.[5, 10-13] As such,
there are numerous methods that have been developed to achieve the desired doping, including;
microwave-assisted polyol;[14-17] electrochemical;[17-19] sonolysis [17, 20] and auto-reduction
deposition.[17, 21] In spite of these recent achievements the goal of a commercially available CNT
electrode has not been realized.

In this paper we present a novel, flexible, 3D CNT composite architecture that has great
potential for a range of electrochemical devices

EXPERIMENT

Carbon Nanotube Growth

3D CNT scaffolds were produced in a 3 stage Atomate tube furnace (Figure 1) utilizing
Iron (III) Tosylate as a pre-cursor catalyst and acetylene as the carbon source. Samples were
prepared by spin casting 2%, 5% or 10%, Fe (III) Tosylate in ethanol onto 4x10cm quartz slides
at 1000rpm. The slides were dried at 100°C for 10min to remove excess solvent and to prevent
catalyst agglomeration. Following drying, the sheets were placed into the furnace between 100
and 400mm from the gas inlet and sealed. Argon was introduced at a flow of 200mL/min with

the furnace held at 100°C for 30min to purge oxygen from the system. The Argon flow was kept constant throughout the CVD experiment as a carrier gas. Reduction of the catalyst was performed between 500-600°C at a flow rate of 10mL/min of H_2 gas. The temperature gradient was further increased to 800-900°C prior to the introduction of acetylene at 20mL/min. Deposition was then performed for varying periods of time ranging from 15 to 60 minutes.[9]

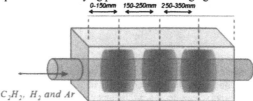

Figure 1. Schematic of the 3-Stage tube furnace utilized for the production of entangled carbon nanotube networks

Nanoparticle Deposition

Composite CNT/Metal NP electrodes were synthesized by a microwave-assisted polyol reduction method. The CNT scaffold was immersed in a solution of 20μL of 8% H_2PtCl_6 ($HAuCl_4$ or $PdCl_2.xH_2O$) (Sigma-Aldrich) in 10mL Ethylene Glycol (EG) and was exposed to 1200W microwave irradiation for 5second periods until reduction was observed to occur.

Carbon Nanotube Growth

Scanning electron microscopy was carried out on a Jeol 6460 with no conductive coating required for imaging. Electrochemical characterization, in the form of cyclic voltammograms (CVs), were tested in a conventional three electrode cell in 0.1M Tetrabutylammonium perchlorate in acetonitrile against an Ag/Ag^+ reference electrode. Capacitances were measured at 0V and power and energy densities were calculated from the cyclic voltammograms.

DISCUSSION

It is well know that CVD of CNTs is highly dependant on a range of variables, including carbon source gas, decomposition temperature, and growth time. Optimization of these parameters for entangled CNT (eCNT) composite architectures has not been examined in detail. Indeed it is theorized that these variables will have an even greater affect in eCNT systems due to the propensity of defects visible in the structure. Previously we have reported the large scale production of unoptimised 3D eCNT architectures.[9]

Initial work on the optimization of these architectures focused on a quantitative study of the effect of catalyst concentration and furnace position. Analysis of the resultant samples via SEM imaging (Figure 2) revealed a cross-dependence on these two variables. Specifically, for the 2% sample, the quality of growth was optimal at the earliest furnace position (150mm) whereas for the 10% sample the final zone (250mm) examined was the optimal position. This position is defined by the emergence of long tubular structures in preference to significantly shorter carbon structures. This variation is explained by considering the roles of the mixture of $H_{2(g)}$ and $C_2H_{2(g)}$ flowing through the furnace. The hydrogen gas is required to activate the nanoparticles for growth, so the more H_2 which the slides are exposed to the more catalyst is available for the growth of CNTs. As such the samples in the earlier zone (150mm) have a greater proportion of activated catalysts than the later zones. The acetylene on the other hand decomposes to produce the carbon radicals which will interact with the activated catalysts, however if no catalyst is available these radicals recombine to form amorphous carbon. As such early in the furnace, at the highest gas mixture concentration, there is a maximum amount of catalyst activated. At this position for the lower catalyst concentrations (2% and 5%) these active catalysts capture an optimal amount of carbon radicals. However for the 10% catalytic sample there are too many active sites capturing radicals not allowing for the formation of the desired tubular architectures. At positions further from the gas inlet (200mm and 250mm) there are less catalytic particles activated which appears to be optimal for the 10% catalyst slide. The decrease in particle activation is detrimental for the 2% catalyst slides as there are excess carbon radicals over these slides depositing undesired amorphous carbon and causing a loss of tubular definition in the samples.

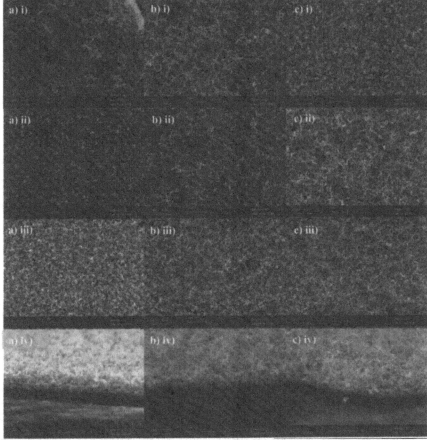

Figure 2. SEM images of eCNT network produced by varying the initial concentration of catalyst (Iron (III) Tosylate) **a)** 2%; **b)** 5%; and **c)** 10% and position from the gas source **i)** 150mm; **ii)** 200mm; and **iii-iv)** 250mm;

The variability in structures, induced by relatively minor changes in the growth conditions, is highlighted by the electrochemical performance of these electrodes. A simple capacitive analysis (Figure 3) reveals that there is a clear difference in stability between the 2% and 10% samples at 150mm. The 2% at150mm sample is stable at more than twice the scan rate of the equivalent 10% sample. This dramatic decrease arises from impedance of the TBAP ions moving through the sample, which in turn is induced by the loss of micro-porosity, observed in the previous micrographs, shown in Figure 3. Increased resistance within the composite system is evident from the CVs which is also reflected in the increase in the RC time constant. Despite this loss of ionic mobility at high scan rates it appears at slow scan rates that the capacitive response of the two materials is quite similar (Table 1), indicating the total available electro-active surface area is fundamentally similar for both samples.

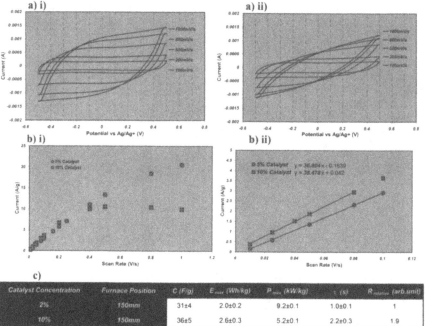

Catalyst Concentration	Furnace Position	C (F/g)	E$_{max}$ (Wh/kg)	P$_{max}$ (kW/kg)	τ (s)	R$_{relative}$ (arb.unit)
2%	150mm	31±4	2.0±0.2	9.2±0.1	1.0±0.1	1
10%	150mm	36±5	2.6±0.3	5.2±0.1	2.2±0.3	1.9

Figure 3. Electrochemical analysis of entangled carbon architectures; **a)** cyclic voltammetry in 0.01M TBAP/ACN; **i)** 2% catalyst @ 150mm; **ii)** 10% @150mm; **b)** capacitance calculations; **i)** at the full range of scan rates; **ii)** within the full linear region at slower scan rates; and **c)** tabulated comparison of capacitive, energy and power densities determined from the cyclic voltammograms.

Deposition of conductive nanoparticles onto the highly porous structures allows for the tuning of these generic scaffolds towards specific applications whilst also increasing the available surface area. A microwave-assisted polyol reduction was developed for use on the eCNT architecture producing nanoparticles of sub 20nm for Gold, Palladium and Platinum. This method showed excellent deposition of nanoparticles onto the MWNTs (Figure 4) (only

Platinum nanoparticles shown). Figures 4a and 4b show the emergence of a peak at ≈37° in the XRD diffraction pattern corresponding to platinum deposition onto the composite structure. Photon correlation spectroscopy (Figure 4d) of the Au, Pd and Pt nanoparticle solutions confirmed the production of sub 20nm particles for Au and Pd, and sub 10nm particles were measured for the Pt samples.

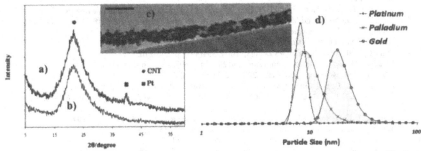

Figure 4. XRD spectra of **a)** NanoWeb architecture coated with platinum; and **b)** untreated NanoWeb architechture; **c)** TEM image of MWNT coated with Platinum nanoparticles (scale bar is 50nm); and **d)** photon correlation spectra for Platinum, Palladium and Gold nanoparticles.

The exceptional loading of Platinum nanoparticles observed arises due to the combination of intense heating of the CNTs causing localized metal reduction at the CNT/electrolyte interface and the concurrent formation of defect sites having dangling bonds which the particles can attach to. The effect of the deposition on the conductivity of the architectures is minimal with the resistance approximately 47±2 Ω/square. The variation in the capacitance of these electrodes, however, is extremely dramatic with the capacitance determined to be as high as 300±20 F/g.

CONCLUSIONS

The growth of 3D eCNT architectures via CVD is highly dependant on a number of variables. Furthermore these variables are not independent of one another, rather there is a clear interaction between these variables which must be accounted for before the ability to 'design' an electrode is attainable. The electrochemical response of the architectures is quite good at 36±F/g and very stable at scan rates up to 500mV/s. The addition of nanoparticles to the eCNT structure increased the electro-active surface area resulting in a 10fold increase in the pure capacitive response of these composite electrodes.

ACKNOWLEDGMENTS

The authors acknowledge ongoing support of the Arustralian Research Council for funding through the Discovery Projects (DP0877348) and Centre of Excellence Program (CE0561616). AIM acknowledges an ARC QEII Research Fellowship (DP0558091).

REFERENCES

1. Kimizuka, O., et al., *Carbon*, 2008. **46(14)**: p. 1999-2001.
2. Bordjiba, T., M. Mohamedi, and L.H. Dao, *Journal of The Electrochemical Society*, 2008. **155(2)**: p. A115-A124.
3. Bordjiba, T., M. Mohamedi, and L.H. Dao, *Advanced Materials*, 2008. **20(4)**: p. 815-819.
4. Balasubramanian, K. and M. Burghard, *Journal of Materials Chemistry*, 2008. **18(26)**: p. 3071-3083.
5. Gooding, J.J., *Electrochimica Acta*, 2005. **50(15)**: p. 3049-3060.
6. Riccardis, M.F.D., et al., *Carbon*, 2006. **44(4)**: p. 671-674.
7. Sun, X., et al., *Chemical Physics Letters*, 2004. **394(4-6)**: p. 266-270.
8. Paradise, M. and T. Goswami, *Materials and Design*, 2007. **28(5)**: p. 1477-1489.
9. Chen, J., et al., *Advanced Materials*, 2008. **20(3)**: p. 566-570.
10. Chang, L. and C. Hui-Ming, *Journal of Physics D: Applied Physics*, 2005. **(14)**: p. R231.
11. Thostenson, E.T., Z. Ren, and T.-W. Chou, *Composites Science and Technology*, 2001. **61(13)**: p. 1899-1912.
12. Kim, H.-T., J.-K. Lee, and J. Kim, *Journal of Power Sources*, 2008. **180(1)**: p. 191-194.
13. Wang, X., et al., *Journal of Power Sources*, 2006. **158(1)**: p. 154-159.
14. Raghuveer, M.S., et al., *Chem. Mater.*, 2006. **18(6)**: p. 1390-1393.
15. Bonet, F., et al., *Nanostructured Materials*, 1999. **11(8)**: p. 1277-1284.
16. Chen, W.-X., J.Y. Lee, and Z. Liu, *Materials Letters*, 2004. **58(25)**: p. 3166-3169.
17. Georgakilas, V., et al., *Journal of Materials Chemistry*, 2007. **17**: p. 2679-2694.
18. Zou, Y., et al., *Biosensors and Bioelectronics*, 2008. **23(7)**: p. 1010-1016.
19. He, Z., et al., *Diamond and Related Materials*, 2004. **13(10)**: p. 1764-1770.
20. Wang, Z.-C., et al., *Journal of Solid State Electrochemistry*, DOI: 10.1007/s10008-008-0558-7.
21. Chen, W., et al., *J. Am. Chem. Soc.*, 2006. **128(10)**: p. 3136-3137.

Mater. Res. Soc. Symp. Proc. Vol. 1142 © 2009 Materials Research Society 1142-JJ04-04

The Role of Defects in Carbon Nanostructures Probed through Ion Implantation and Electrochemistry

Mark Hoefer[1], Jeff Nichols[1], and Prabhakar Bandaru[1]
[1]Materials Science and Engineering, U. of California, San Diego, La Jolla, Ca, 92093-0418, USA

ABSTRACT

As carbon nanotubes (CNTs) inevitably contain defects, an understanding of their effect on the electrochemical behavior is crucial. We consider, through Cyclic Voltammetry and Raman Spectroscopy the influence of both intrinsic and extrinsically introduced defects. Bamboo and hollow multi-walled carbon nanotube morphologies provided examples of the former while the controlled addition of Argon and Hydrogen ions was used for studying extrinsic defects. We show that the electrocatalytic response of the hollow type CNTs can be tailored significantly, while bamboo type CNTs have innately high reactive site densities and are less amenable to modification. Argon irradiation also differs greatly from that of Hydrogen irradiation. CNT irradiation with Argon appears to positively charge CNTs, while Hydrogen irradiation neutralizes defects further allowing for the tuning of CNT defect density. The work has implications in the design of nanotube and nanowire based chemical sensors.

INTRODUCTION

The postulated fast electron transfer kinetics[1], related to the large surface area/ volume ratios of carbon nanotubes (CNTs), could be useful for the development of increased sensitivity, CNT based, electrode materials, electrochemical sensors, supercapacitors etc.,[2, 3]. In this context, it has been pointed out that the electrocatalytic behavior along the length of the CNTs would be similar to the basal planes of graphite, while the ends correspond to the edge planes[4, 5] and could be influenced by dangling bonds[6], as in edge-plane graphite. The latter corresponds to a large defect density, which could be profitably used for the enhanced sensitivity. While the nature and extent of the defects in CNTs can also be altered through the addition of functional moieties[7-9], in this study, we investigate the influence of external irradiation on nanotube behavior. We show that Ar and H ions can be used to systematically tune the electrochemistry of CNTs.

We investigated the tunability characteristics in both multi-walled, hollow-core carbon nanotubes (HCNTs), and bamboo-type carbon nanotubes (BCNTs). Structural modification, through defects, influences the predicted performance of CNTs for high efficiency electrodes[6, 10, 11]. Such sensitivity was investigated through Cyclic Voltammetry (CV) and Raman Spectroscopy. Consequently, non-ideal electrode behavior, linked to irreversible electron transfer/adsorption processes[12] were monitored.

EXPERIMENTAL DETAILS

Both the hollow core (HCNTs) and bamboo (BCNT) carbon nanotube morphologies were grown via thermal chemical vapor deposition (CVD) on Si substrates, aided by 5 nm thick Fe catalyst[13, 14] to form vertically aligned arrays. The BCNTs were synthesized at 850°C with a gas feedstock comprising 100mL/min of benzene, 500 sccm of Ar, and 200 sccm of NH3. The HCNTs were grown at 615°C using acetylene gas (50 sccm, for 1 min) and 500 sccm Ar. Subsequent to growth, the CNT samples were subject to Argon and Hydrogen irradiation in with increasing Ar/H exposure. Raman spectroscopy was performed on the untreated and Ar/H exposed samples using a 514.5 nm Argon ion laser (Renishaw) at a power of 1.49 mW. From the spectographs the crystallite size and relative defect density were determined.

The irradiated CNTs were also investigated for their electrochemical properties, through amperometric response, by placing them as working electrodes in CV experiments. (Figure 1a)

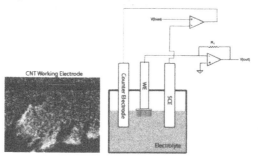

Figure 1: (a) Arrays of CNTs, (b) Schematic of electrochemical experiment.

Since the inter-nanotube distance was ~500 nm, the diffusion layer of each CNT composing the electrode overlaps yielding macro-electrode like behavior[5]. A standard three electrode setup (Figure 1b), in a 1M KCl supporting electrolyte solution containing various concentrations (1-10 mM) of $K_3Fe(CN)_6$, was employed with a (i) HCNT/BCNT working electrode, (ii) platinum wire counter electrode, and a (iii) saturated calomel reference electrode (SCE). It was verified, through control experiments, that neither the silicon nor the epoxy displayed any electrochemical activity in the voltage window (-0.35 V to 0.8 V) used for the CV. We also took care to eliminate the possibility of hexacyanoferrate (HCF) complex adsorbate formation on the electrodes15, which could affect electrochemical kinetics, through the choice of the voltage scan range (-0.4 to 0.8 V) and using freshly prepared (< 2 hours old) $Fe(CN)_6^{4-}/Fe(CN)_6^{3-}$ solutions. The experiments were carried out at different scan rates (1 mV/sec – 1 V/sec) and at room temperature.

EXPERIMENTAL RESULTS AND DISCUSSION

Raman Spectroscopy characterization of HCNTs and BCNTs:

As the CNTs are exposed to irradiation, the structural changes were represented by the changes in the Raman G-, D-peak intensities and areas. (Figure 2)

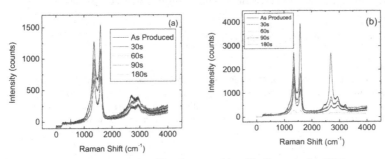

Figure 2: Raman spectrographs of bamboo type(a) and hollow type(b) CNTs.

With increasing Ar irradiation, the G-peak frequency was up-shifted with increased disorder (represented by the FWHM). This phenomenon suggests that Ar ions are being intercalated into the graphene planes forming acceptor-like defects corresponding to a contraction of the inter-planar bond lengths, spanning a wide distribution of energies. A larger increase in the line intensity ratio, $(I_D/I_G)_L$, for the HCNTs compared to the BCNTs is also observed(Figure 3a).

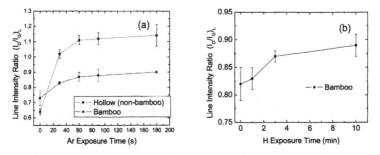

Figure 3: Increase in line intensity for argon irradiation(a) and H irradiation(b).

Similarly, the integrated area ratio $(I_D/I_G)_A$ for HCNTs shows approximately a 30% increase (from 1.05 to 1.40), while exhibiting a much smaller range of variation (~10%)

for the BCNTs. While the G-peak and $\Delta\omega_G$ for both HCNTs and BCNTs approach similar values with increased argon exposure, indicating a preponderance of defects, a larger increase in $\Delta\omega_G$ for HCNTs, compared to BCNTs, from the untreated forms indicates that the HCNTs are more available for defect engineering. While argon does appear to be intercalated into BCNTs, the original defects appear to dominate.

With increasing H irradiation no peak shifts were observed. Changes in the line intensity ratio (from .82 to .89 in Figure 3b) and the $\Delta\omega$'s (~25% for the G-peak and ~50% for the D-peak as shown in Table I) however were observed, indicating that the relative defect density in the BCNTs is increased, but with a variety of energies. This phenomenon is also believed to result from the termination of dangling bonds16.

Cyclic Voltammetry (CV) of HCNTs and BCNTs:

Argon Irradiation:
In the CV characterization (Figure 4), it was seen that while the ratios of the cathodic and anodic current densities, i.e., $|i_{pc}/i_{pa}|$ was approximately unity for both types of nanotubes with Ar irradiation, the irreversibility could be better indicated by ΔE_p and deviation from the ideal value of 59 mV (for a one-electron redox reaction).

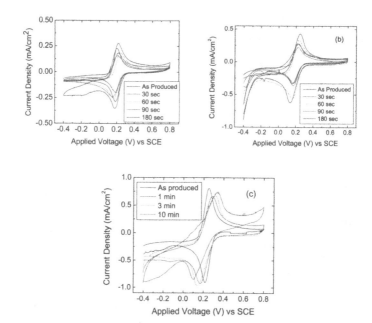

Figure 4: CV curves for Argon irradiated (a)HCNTs and (b) BCNTs and Hydrogen irradiated (c) BCNTs.

20

It was seen that the initial ΔE_p for HCNTs was ~ 62.5 mV and for the BCNTs ~ 45.5 mV. With increased argon exposure, the ΔE_p exhibited a much larger variation for the HCNTs (increasing from 62.5 mV to 118 mV, ~ 90%) compared to the BCNTs (which was enhanced by ~ 40%, from 45.5 mV to 65 mV). Consequently, both HCNTs and BCNTs showed quasi-reversible/diminished electron transfer kinetics with increased argon exposure, with a greater influence of the defects for the former. As we posited earlier that the intrinsic morphological defects dominate BCNT characteristics, and HCNTs are more amenable to defect introduction, the variation of the ΔE_p can be justified. The individual variation of E_{pc} / E_{pa} has a larger range in the HCNTs (Table VII), while the current densities, which are more a function of the number of defect sites, and not their nature, were relatively unchanged in either nanotube variety.

The hypothesis of argon exposure positively charging the nanotubes[32] (Figure 5, indicates the relevant mechanisms) made through Raman spectroscopy investigations, was borne out in CV characterization by the increasingly positive (/negative) change of E_{pc} (/ E_{pa}) for charge transfer from the CNT electrode to the redox couple.

Hydrogen Irradiation:

BCNTs under H irradiation experienced both a loss of reversible electron transfer kinetics and a decrease in the electron transfer rate constant. Upon irradiation $|i_{pc}/i_{pa}|$ decreased to .77 and ΔE_p increased to ~237mV indicating a large deviation away from ideality. As postulated, this behavior is expected with the loss of reactive sites due to dangling bond quenching with H irradiation.

CONCLUSIONS

We have shown that exposure to argon irradiation can influence the charge state of both bamboo morphology and hollow CNTs through Raman Spectroscopy and concomitant Cyclic Voltammetry characterization. It is plausible that argon leads to intercalation, where it abstracts electrons from the CNT, creating acceptor like defects. It was also seen that the initial structural state, can limit the relative amount of charge and defects that could be introduced. While bamboo morphology nanotubes had an intrinsically higher defect concentration and exhibit a smaller variation on argon exposure from reversible electron transfer kinetics, hollow variety nanotubes are more amenable to modification, with greater variation. Furthermore, we have demonstrated that H irradiation also allows the electrochemical response of CNTs to be tuned. In the case of H irradiation however, both the reversibility and electron transfer kinetics are degraded due to the loss of reactive sites(dangling bonds).

ACKNOWLEDGMENTS

We gratefully acknowledge support from the National Science Foundation (Grants ECS-05-08514) and the Office of Naval Research (Award number N00014-06-1-0234). We are thankful to Professor Frank Talke's group at the Center for Magnetic Recording Research (CMRR) at UC, San Diego for help with the Raman spectroscopy.

REFERENCES

1. Nugent, J. M., Santhanam, K.S.V., Rubio, A., and Ajayan, P.M., *Nanolett.*, 1, 87-91 (2001).
2. Baughman, R. H.; Zakhidov, A. A.; de Heer, W. A., *Science*, 297, 787 (2002).
3. Zhao, Q., Gan, Z., and Zhuang, Q., *Electroanalysis*, 14, 1609-1613 (2002).
4. Banks, C.; Moore, R.; Davies, T.; Compton, R., *Chem. Comm.*, 16, 1804-1805 (2004).
5. Banks, C. E., Davies, T.J., Wildgoose, G.G., and Compton, R.G., *Chem. Comm.*, 7, 829-841 (2005).
6. Chou, A. B., T., Singh, N.; Gooding, J., *Chem. Comm.*, 7, 842-844 (2005).
7. Wang, J., Kawde, A.; Jan, M. R., Biosensors and Bioelectronics, 20, 995-1000 (2004).
8. Zeng, Y. L.; Huang, Y. F., Jiang, J. H.; Zhang, X. B.; Tang, C. R.; Shen, G. L.; Yu, R. Q., *Electrochem. Comm.*, 9, 185-190 (2007).
9. Strano, M. S., Dyke, C. A.; Ursey, M. L.; Barone, P. W.; Allen, M. J.; Shan, H.; Kittrell, C., Hauge, R. H.; Tour, J. M.; Smalley, R. E., *Science*, 301, 1519-1522 (2003).
10. Frackowiak, E., Metenier, K., Bertagna, V., and Beguin, F., *Appl. Phys. Lett.*, 77, 2421-2423 (2000).
11. Wang, J., Sun, X.; Cai, X.; Lei, Y.; Song, L.; Xie, S., *Electrochem. and Solid-State Lett.*, 10, J58-J60 (2007).
12. Bard, A. J., Faulkner, L. R., Electrochemical Methods: Fundamentals and Applications. 2 ed.; John Wiley: New York, 2001.
13. Deck, C. P., Vecchio, K., *Carbon*, 43, 2608-2617 (2005).
14. Nichols, J., Deck, C. P.; Saito, H.; Bandaru, P. R., *J. Appl. Phys.*, 102, 064306 (2007).
15. Pharr, C. M., Griffiths, P. R., *Anal. Chem.*, 69, 4673-4679 (1997).
16. Wang, W. K., Lin, X. W., Mesleh, M., Jarrold, M. F., Dravid, V. P., Ketterson, J. B., Chang, R. P. H., *J. Mater. Res.*, 10, 1977-1983 (1995).

Carbon Nanotube Synthesis
and Characterization

Mater. Res. Soc. Symp. Proc. Vol. 1142 © 2009 Materials Research Society   1142-JJ05-05

Higher Yield of Short Multiwall Carbon Nanotubes by Catalytic Growth

V. Z. Mordkovich, A. R. Karaeva, S. V. Zaglyadova, I. A. Maslov, A. K. Don, E. B. Mitberg and D. N. Kharitonov
United Research Development Centre, Leninsky pr. 55/1, Bldg. 2, Moscow 119333, Russia

ABSTRACT

This work is dedicated to formulation of growth conditions for higher yield CNT growth on relatively simple and cheap Fe/Al_2O_3 catalysts. Fe/Al_2O_3 catalysts of seemingly identical composition are reported to give wildly varying CNT yields. Different crystalline forms of Al_2O_3 (γ-Al_2O_3 or δ-Al_2O_3 or θ-Al_2O_3 or α-Al_2O_3) were used in this work for catalysts preparation, Fe content was varied, too. The best results were reached for the catalyst 10 % Fe/δ-Al_2O_3 prepared with the use of ethanol solution of $Fe(NO_3)_3 \cdot 9H_2O$. The yield of 19100 g/g_{Fe} was reached at 650°C in 3-hour experiment with the use of benzene-saturated hydrogen gas (saturation temperature 20°C) as a precursor. Characterization with TEM, TGA and XRD showed that the product is dominated by short multiwall CNT with relatively narrow size distribution (20-50 nm diameter), 500-600 nm length) and nearly cylindrical layer stacking. No non-CNT carbon was observed by TEM, with this observation supported by TGA manifestation of a single peak of CNT oxidation at 615°C. The results witness very high yield of carbon nanotubes suitable as filler material for composites. The absence of non-CNT carbon in the product of catalytic growth makes final purification much easier.

INTRODUCTION

The research works in the field of carbon nanotube (CNT) catalytic growth are mostly devoted to one of the three major challenges, i.e. (a) selective growth of certain fractions; (b) growth of longer CNT for technological fibers or (c) growth of short CNT with maximum yield. The latter task is most closely related to the emerging industry of multi-ton CNT production for polymer-CNT composites and other applications. The yields of lower than 300 g/g, which were typical for earlier catalytic growth works reported before 2003 are no longer interesting for future development. The future of higher-yield CNT growth is related to satisfaction of three major requirements, i.e. (a) yield higher than 1000 g/g; (b) impurities of non-CNT as low as possible and (c) cheaper catalyst.

While powder and supported metals exhibit the different reactivity and CNT growth characteristics, metals on different supports seem to have also different activity and CNT growth due to different nature and strength of the metal-support interactions. The effects have been manifested in many studies [1-3]. In addition, the material, morphology, and textural properties of the substrate greatly affect the yield and quality of the resulting CNTs. Zeolite supports with catalysts in their nanopores have resulted in significantly higher yields of CNTs with a narrow diameter distribution [4]. Alumina materials are reported to be the better catalyst supports than silica owing to their strong metal-support interaction, which allows the high metal dispersion and, thus, high density of the catalytic sites [5]. A good summary of the effects of metal-support interaction was given by Anderson et al. [6]: it induces the electronic perturbations throughout the metal; it generates the significant differences in the metal morphology and the arrangement of the surface atoms; it exerts an influence of the growth characteristics of the supported metal particles; and it possibly modifies the chemistry of the system. Even the origin of the metal precursors or the catalyst preparation procedures might play a role. Serp et al. [7] found that among iron chloride, iron sulfate and iron carbonyl, $Fe_3(CO)_{12}$ was the best precursor. Li et al. [8] reported one of the highest yields of 600 % while investigating the influence of the partial pressure of acetylene precursor over Fe/SiO_2 catalyst.

This work is dedicated to formulation of growth conditions for higher yield CNT growth on relatively simple and cheap Fe/Al_2O_3 catalysts. It is necessary to note that the Fe/Al_2O_3 catalysts of seemingly identical composition are reported to give wildly varying CNT yields. We are set to confirm that it is possible to reach reproducible, high yields of high quality CNT on these catalysts.

EXPERIMENT

The catalysts were prepared by impregnation of Al_2O_3 powder supports with aqueous or ethanol solutions of $Fe(NO_3)_3 \cdot 9H_2O$ to provide 1, 5, 10 or 20 % Fe content in the support. The impregnated supports were then dried and annealed. The supports were prepared by thermal treatment of boehmite AlO(OH), which was provided by Sasol (product ID Pural SB-1). Depending on the temperature of the treatment we received γ-Al_2O_3 or δ-Al_2O_3 or θ-Al_2O_3 or α-Al_2O_3 (see Fig. 1), which was confirmed by X-ray diffraction.

$$Boehmite \xrightarrow{\;300-500^\circ C\;} \gamma - Al_2O_3 \xrightarrow{\;850^\circ C\;} \delta - Al_2O_3 \xrightarrow{\;1050^\circ C\;}$$

$$\Theta - Al_2O_3 \xrightarrow{\;1200^\circ C\;} \alpha - Al_2O_3$$

Figure 1. The scheme of polymorphic transitions of boehmite.

All the prepared catalysts were activated by hydrogen at 600°C and then the catalytic growth of CNT was carried out in a reaction gas mixture flow at the temperature from 650°C to 1100°C. The reaction gas mixtures contained CH_4, H_2, N_2, benzene or toluene in various ratios. The reaction time was varied from 0.5 hr to 3 hr in order to confirm additionally the reproducibility.

Transmission electron microscopy (TEM), thermal gravimetric analysis (TGA) and X-ray diffraction (XRD) were used for characterization of supports, catalysts and products. The measurements of specific surface (S_{BET}) and porosity of the supports sustained the XRD data on the difference among modifications of Al_2O_3 – see Table I.

Table I. Adsorption characteristics of the synthesized alumina supports.

Modification	S_{BET} (m^2/g)	Volume of pore, (cm^3/g)	Size of pore (Å)
Pural SB-1	224	0.39	67
γ-Al_2O_3	78	0.38	195
δ-Al_2O_3	51	0.42	330
Θ-Al_2O_3	25	0.26	401
α-Al_2O_3	0.9	0.01	59

RESULTS and DISCUSSION

Experiments with all the catalysts led to formation of black powders consisting mostly of short CNT.

We consecutively studied the influence of support structure, reaction temperature, method of Fe impregnation, Fe concentration, feedgas composition and reaction time on the yield of the resulting carbon nanotubes and their structure.

The support structure proved to be very important and influenced the yield critically as shown in Fig. 2, where the results are shown for fixed synthesis duration 1 hr, fixed synthesis temperature 650°C and fixed feedgas composition.

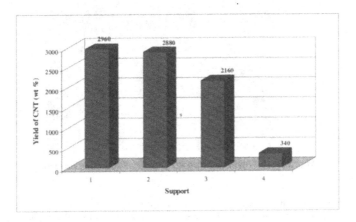

Figure 2. Dependence of the carbon yield on the support type in CNT synthesis using methane-benzene mixture as feed gas: 1 – γ-Al$_2$O$_3$, 2 – δ-Al$_2$O$_3$, 3 – Θ-Al$_2$O$_3$, 4 – α-Al$_2$O$_3$

The use of γ-Al$_2$O$_3$ or δ-Al$_2$O$_3$ forms as supports led to much higher yields in all the experiments, which is consistent with their higher surface area and pore size as can be seen from Table I. The advantage of the γ-Al$_2$O$_3$ over δ-Al$_2$O$_3$ in the CNT yield is never strong, but the reproducibility of δ-Al$_2$O$_3$–supported catalysts proved better, so all the rest of our experiments were done with the use of δ-Al$_2$O$_3$.

The structure of produced nanotubes did not vary much within the parameters of our experiments. A typical structure is shown in Fig. 3. The length of CNT was up to 1.5 μm, the outer diameter up to 25 nm, inner diameter from 5 to 10 nm, the wall thickness 10 to 20 graphene sheets. The layout of the graphene sheets was predominantly cylindrical though with many inclinations and defects. TGA manifested a single peak of CNT oxidation at 615°C. Combination of TEM and TGA data suggest that the samples did not contain traceable amounts of non-nanotube carbon. The only exception was observed at the reaction temperatures over 900°C, where non-catalytic deposition of pyrolytic carbon was registered.

The influence of the reaction temperature was different in two temperature ranges, i.e. in the range 650 to 850°C we observed a decrease in the carbon yield without strong influence on the CNT structure, while in the range 900 to 1100°C we observed a steep decrease in the carbon yield caused by intensive non-catalytic deposition of pyrolytic carbon. Such early manifestation of non-catalytic deposition is explained by high carbon concentration in our feedgas compositions as will be shown below.

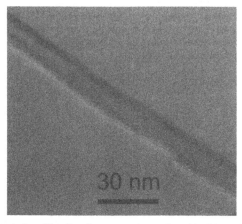

Figure 3. TEM of a typical multiwall carbon nanotube produced in this work.

It was shown that the reaction yield is influenced by a solvent used for impregnation solution preparation. The best results were achieved by using ethanol solution of aqueous ferric nitrate.

The concentration of Fe as an active component proved to be important as shown in Fig. 4. It can be seen that the dependence of the carbon yield on the content of active component (Fe) in the catalyst reached maximum at Fe concentration of 10 % wt. per support. Probably, the saturation of the support surface with metal iron particles takes place at this content of Fe. The further decrease of carbon yield with the increase of Fe content up to 20% wt. is connected with agglomeration of the metal particles on the support surface followed by the decrease of the quantity of the active centers. Anomalously low values of the carbon yield at low concentration of Fe (1 и 5 %), are probably due to the Fe-alumina interactions at the synthesis temperature with formation of spinels and spinel-based structures, which can block some of the active component.

The feedgas composition was varied by changing the methane/hydrogen ratio and by introducing certain amounts of benzene or toluene by saturating the gas flow at fixed temperature. It was shown that the gas flow saturation by benzene at 20°C led to much higher yields without harm to the CNT structure. The dilution of the gas mixture by hydrogen led to higher yields as well. As shown in the Fig. 5, the highest yield was reached at "infinite" dilution i.e. where the main source of carbon in the feedgas was benzene. The data in Figs 4, 5 are shown for the reaction time 3 hrs. Our measurements showed that the yield grew linearly with time up to the reaction time 3 hrs, which was selected as an optimum reaction duration.

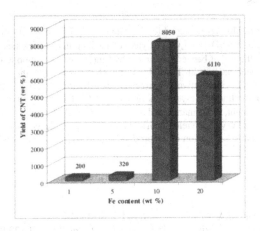

Figure 4. The influence of iron content in the catalyst on the yield of carbon nanotubes. (reaction conditions: T=650°C, $CH_4:H_2$ = 2:1, $T_{benzene}$=30°C, τ=3h).

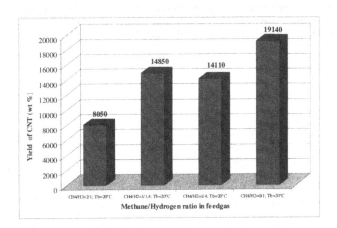

Figure 5. The influence of the gas mixture composition on the carbon yield.

CONCLUSIONS

The best results were reached for the catalyst 10 % Fe/δ-Al$_2$O$_3$ prepared with the use of ethanol solution of Fe(NO$_3$)$_3$·9H$_2$O. The yield of 19100 g/g$_{Fe}$ was reached at 650°C in 3-hour experiment with the use of benzene-saturated hydrogen gas (saturation temperature 20°C) as a precursor. Characterization with TEM, TGA and XRD showed that the product is dominated by short multiwall CNT with relatively narrow size distribution (20-50 nm diameter. 500-600 nm length) and nearly cylindrical layer stacking. No non-CNT carbon was observed. Carbon content was 97 to 98 % weight, TGA manifested a single peak of CNT oxidation at 615°C.
The results witness very high yield of carbon nanotubes suitable as filler material for composites. The absence of non-CNT carbon in the product of catalytic growth makes final purification much easier.

REFERENCES

1. C. Park, and M.A. Keane, *J. Catal.* **221**, 386 (2004).
2. S. Takenaka, M. Serizawa, and K. Otsuka, *J. Catal.* **222**, 520 (2004).
3. T. de los Arcos, M.G. Garnier, J.W. Seo, P. Oelhafen, V. Thommen, and D. Mathys, *J. Phys. Chem.* **B108**, 7728 (2004).
4. K. Hernadi, et al., *Zeolites* **17** (5-6), 416 (1996).
5. N. Nagaraju, , et al., *J. Mol. Catal. A* **181** (1-2), 57 (2002).
6. P. E. Anderson, and N.M. Rodriguez, *Chem. Mater.* **12**, 823 (2000).
7. J. L. Figueiredo, and Ph. Serp, "Optimizing growth conditions for carbon filaments and vapor-grown carbon fibers", *Carbon Filament and Nanotubes: Common Origins, Differing Applications*, ed. L. P. Biro (Kluwer, 2001) pp. 111-120.
8. W. Z. Li, J. G. Wen, Y. Tu, and Z. F. Ren, *Appl. Phys. A* **73**, 259 (2001).

Mater. Res. Soc. Symp. Proc. Vol. 1142 © 2009 Materials Research Society   1142-JJ05-07

Enhancement of Single-Walled Carbon Nanotube Formation Using Aluminum Oxide Buffer Layer in Alcohol Gas Source Method

Takahiro Maruyama, Tomoyuki Shiraiwa, Kuninori Sato, Shigeya Naritsuka
Department of Materials Science and Engineering, Meijo University, 1-501 Shiogamaguchi, Tempaku, Naogya 469-8502, Japan

ABSTRACT

Aluminum oxide (Al_2O_x) buffer layers were employed in single-walled carbon nanotube (SWNT) growth using an alcohol gas source and Co catalyst. By optimizing the thickness of deposited Al, the SWNT yield reached its maximum when Al thickness was 30 nm. Transmission electron microscopy (TEM) revealed that Co particle size was strongly dependent on Al thickness. Dense Co particles about 2-4 nm in diameter were formed at the optimal Al thickness, which provided a significant enhancement to yield. In addition, Co diffusion into the Al_2O_x layer was also observed, which would reduce the yield at higher Al_2O_x thickness.

INTRODUCTION

Since the discovery by Iijima [1], carbon nanotubes (CNTs) have been highly anticipated for use in various nanoelectronics devices because of their excellent electronic properties. Recently, we have reported CNT growth by a gas source method in an ultra-high vacuum (UHV) chamber using ethanol gas with a Co catalyst [2]. This growth technique enables CNT growth in a high vacuum, which is useful for investigation of the growth mechanism. In addition, CNT growth in a high vacuum reduced the growth temperature below 400°C. In spite of these advantages, the yields of CNTs grown by the gas source method have been small, because of a low carbon supply.

To enhance CNT yield, several buffer layers, such as Al_2O_x, TiN, and TiO_2, have been attempted in chemical vapor deposition (CVD) growth, using a hydrocarbon source with Fe or Ni catalysts [3-5]. Among these, Al_2O_x has proven effective, and is widely used for CNT growth. In this study, we utilized an Al_2O_x layer in CNT growth from an ethanol gas source, attaining a drastic enhancement of yield. We also characterized both the Al_2O_x layers and Co particles in detail, further clarifying the yield enhancement mechanism.

EXPERIMENT

Co/Al$_2$O$_x$ catalysts were formed on SiO$_2$(100 nm)/Si substrates. The Al$_2$O$_x$ layers were formed by depositing Al on the substrates with arc-pulsed plasma in a UHV chamber, followed by exposure to air. Co catalysts were deposited on the Al$_2$O$_x$ layers at room temperature by electron beam (EB) deposition. The nominal thickness of the Al was varied from 0 to 60 nm, whereas the Co thickness was set to a constant 0.1 nm.

Ethanol gas was then supplied to the substrate surface through a stainless steel nozzle for 10 min to grow CNTs. The sample temperature was monitored by a pyrometer, and maintained at 700°C during CNT growth. The supply of ethanol gas was controlled by monitoring the ambient pressure, and was kept at 1×10^{-1} Pa. To allow TEM observation, Mo mesh grids with carbon support films (Okenshoji, Mo200) were also used as substrates for CNT growth, under the same growth conditions as the SiO$_2$/Si substrates.

The resulting CNTs were characterized by Raman spectroscopy (Jobin Yvon; Ramanor T6400), and TEM (JEOL; JEM-3200). The excitation wavelengths in the Raman measurements were 514.5 nm. The Al$_2$O$_x$ buffer layers were characterized by secondary ion mass spectrometry (SIMS; PHI ADEPT1010).

RESULTS AND DISCUSSION

Figure 1 shows the Raman spectra of the radial breathing mode (RBM) and the high-energy mode for the CNTs grown on Co/Al$_2$O$_x$/SiO$_2$/Si substrates at various Al$_2$O$_x$ layer thicknesses. All spectra were measured under the same geometry, and the laser intensity was calibrated on the sample surface. Figure 1(a) shows several peaks at 190 cm^{-1} and around 270 cm^{-1} in the RBM region of each spectrum. Since the diameter of a CNT, d, is related to the Raman shift, ω, in the RBM as d (nm) = 248/ω (cm^{-1}) [6], the diameter of CNTs leading to the 190 cm^{-1} peaks was about 1.3 nm, and those generating the 270 cm^{-1} peaks ranged from 0.9 to 1.0 nm. As the thickness of the Al layer increased, the RBM peak intensities increased, reaching a maximum at an Al thickness of 30 nm, then decreasing when Al thickness was further increased to 60 nm. G band peak intensity also increased with Al thickness, reaching a maximum at an Al thickness of 30 nm, as shown in Fig. 1(b). Regardless of Al thickness, the D band intensity was fairly low, and the G/D ratio of the CNTs was about 48 at an Al thickness of 30 nm, where the G band reached its maximum. The Raman results demonstrate that the quality of grown CNTs was improved by the Al$_2$O$_x$ buffer layers, and the optimal Al thickness, in regard to both yield and quality, was about 30 nm.

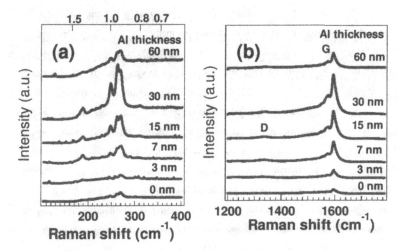

Figure 1. (a) Raman spectra of RBM region for CNTs grown on Co/Al$_2$O$_x$/SiO$_2$/Si substrates at various Al$_2$O$_x$ layer thicknesses. (b) Raman spectra of high frequency regions for the samples shown in (a).

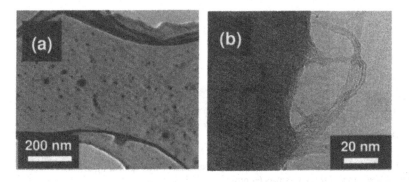

Figure 2. (a) and (b) TEM images of Al$_2$O$_x$ buffer layer formed on the grids after CNT growth. The deposited Al thickness was (a) 3 and (b) 30 nm.

To investigate the CNTs and catalysts in detail, CNTs were grown on TEM grids with Al$_2$O$_x$ buffer layers, and then observed by TEM. Figures 2 (a) and (b) show TEM images of the Al$_2$O$_x$ layers formed by 3 and 30 nm Al deposition, respectively. In Fig. 2(b), a bundle of single-walled CNTs (SWNTs) was observed with diameters of approximately 1-2 nm, which was consistent with the Raman spectra (Fig. 1(a)). In addition, many dark spots were observed in both Al$_2$O$_x$ buffer layers. For an Al thickness of 3 nm, the spot diameters were widely distributed from several nm to several tens of nm. From electron energy loss spectroscopy (EELS), these spots were determined to be composed of Co (not shown). On the other hand, when the Al thickness was 30 nm, the diameter of the majority Co particles was distributed between 2 and 3 nm, and particles of more than 5 nm in diameter were not observed. Catalytic particle size is strongly related to the SWNT diameter with the former generally being smaller than the latter [7]. Because of this, the Co particles formed on 30 nm Al were suitable for growth of SWNTs with diameters below 2 nm. Therefore, it is considered that the increase of SWNT yield was caused by the reduction of Co particle size through optimization of Al thickness. Recently, H. Ohno et al. reported SEM observations that showed that an alumina buffer layer tended to produce highly-dense nanosized Co particles [8]. Our TEM results show direct evidence that the Al$_2$O$_x$ layer produced Co catalyst particles of 2-3 nm in diameter, which are favorable for high efficient SWNT growth.

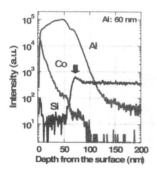

Figure 3. SIMS depth profile of Co/Al$_2$O$_x$/SiO$_2$/Si substrates after annealing at 700 °C for 10 min, when the Al thickness was 60 nm.

Interestingly, CNT yield is strongly dependent on Al thickness, and the yield decreases when the Al_2O_x layer becomes too thick. This indicates that the enhancement of CNT yield is caused not only by surface reaction, but also by bulk related processes. To investigate elemental concentration changes within the Al_2O_x layers, SIMS depth profiles were generated for $Co/Al_2O_x/SiO_2/Si$ substrates after annealing at 700°C (Fig. 3). The nominal thicknesses of Co and Al were 0.1 and 60 nm, respectively. The depth was calibrated by assigning the Al_2O_x/SiO_2 interface to 60 nm. The concentrations of Al, Co and Si are shown in the profile, where the concentration changes are not abrupt at the Al_2O_x/SiO_2 interface, because of both inhomogeneous surface roughness and SIMS resolution limitations. Nevertheless, the Co concentration clearly showed the appearance of an additional peak at the Al_2O_x/SiO_2 interface [arrow in Fig. 3], which was not observed before annealing (not shown). This indicates that, after annealing, Co diffused from the surface into the Al_2O_x layer, leading to Co accumulation at the interface. We believe that for samples with 60 nm thick Al, this diffusion reduced the Co concentration on the Al_2O_x surface, thus lowering CNT yield. Therefore, it is important to optimize Al_2O_x thickness, taking into account both the growth temperature and Co thickness.

CONCLUSIONS

We have demonstrated that Al_2O_x buffer layers caused the Co catalyst to assume nanosized particles (2-3 nm in diameter), thereby enhancing SWNT growth. By including an Al_2O_x layer, the SWNT yield was increased. Also, Co diffusion towards the Al_2O_x/SiO_2 interface occurred before CNT growth, which would reduce the yield at high Al_2O_x buffer thickness.

ACKNOWLEDGMENTS

This work was partially supported by JSPS, the Grant-in-aid for Scientific Research (C) 17560015. We are grateful to Prof. Nishi, Dr. Oishi and Mr. Numao of the Institute of Molecular Science (IMS), Okazaki, for providing the TEM facility under the Nanotechnology Network Project supported by the Ministry of Education, Culture, Sports, Science and Technology (MEXT).

REFERENCES

1. S. Iijima, Nature **354,** 56 (1991).

2. K. Tanioku, T. Maruyama and S. Naritsuka, Diamond Relat. Mater. **17**, 589 (2008).

3. H. Hongo, F. Nihey, T. Ichihashi, Y. Ochiai, M. Yudasaka and S. Iijima, Chem, Phys. Lett. **380,** 158 (2003).

4. T. Acros, M. G. Garnier, P. Oelhafen, D. Mathys, J. W. Seo, C. Domingo, J. V. G. Ramos and S. S. Cortes, Carbon **42,** 187 (2004).

5. S. Noda, K. Hasegawa, H. Sugime, K. Kakehi, Z. Zhang, S. Maruyama and Y. Yamaguchi, Jpn. J. Appl. Phys. **46,** L399 (2007).

6. A. Jorio, R. Saito, J.H. Hahner, C.M. Liever, M. Hunter, T. McClure, G. Dresselhaus and M. S. Dresselhaus, Phys. Rev. Lett. **86,** 1118 (2001).

7. G.-H. Jeong, S. Suzuki, Y. Kobayashi, A. Yamazaki, H. Yoshimura and Y. Homma, J. Appl. Phys, **98,** 124311 (2005).

8. H. Ohno, D. Takagi, K. Yamada, S. Chiasi, A. Tokura and Y. Homma, Jpn. J. Appl. Phys. **47,** 1956 (2008).

Mater. Res. Soc. Symp. Proc. Vol. 1142 © 2009 Materials Research Society   1142-JJ05-09

Carbon Nanotubes Growth on Calcium Carbonate Supported Molybdenum-Transition Metal Catalysts

Zhongrui Li,[*] Enkeleda Dervishi, Yang Xu, Viney Saini, Meena Mahmood, Olumide Dereck Oshin, Alexandru S. Biris[*]
Nanotechnology Center and Applied Science, University of Arkansas at Little Rock, Arkansas, 72204
Alexandru R. Biris, Dan Lupu,
National Institute for Research and Development of Isotopic and Molecular Technologies, P.O. Box 700, R-400293 Cluj-Napoca, Romania

ABSTRACT

A comparison of different catalyst systems (Ni, Co, Fe/Mo nanoparticles supported on $CaCO_3$) has been performed in order to optimize the carbon nanotube (CNT) growth. The influences of the reaction temperature, metal loading and carbon source on the synthesis of CNTs were systematically investigated. Dense CNT networks have been synthesized by thermal chemical vapor deposition (CVD) of acetylene at 720 °C using the Co-Mo/$CaCO_3$ catalyst. The dependence of the CNT growth on the most important parameters was discussed exemplarily on the Co catalyst system. Based on the experimental observations, a phenomenological growth model for CVD synthesis of CNTs was proposed. The synergy effect of Mo and active metals was also discussed.

1. INTRODUCTION

As the carbon nanotube (CNT) applications continuously increase, CNTs of high quality in large quantities are demanded. Catalytic chemical vapor deposition (cCVD) method offers the possibility to produce the large-scale and high-quality production of CNTs at relatively low cost, and the growth of carbon nanotubes can be controlled by adjusting the reaction conditions and choosing proper catalysts. In the cCVD method, CNTs are produced from the thermal decomposition of the carbon-containing molecules on desirable metal catalysts (commonly Fe, Co, and Ni).[1,2] Strong adhesion between CNTs and the catalytic metal clusters is a requirement for CNT growth. This adhesion must be larger than the energy gained by the CNT when carbon dangling bonds form a cap. The catalyst composition (controlled by its preparation method) affects the efficiency and the selectivity of the catalysts towards the synthesis of desired CNTs. Using thermal cCVD, Co and Fe catalysts generally tend to form hollow and well graphitized nanotubes whereas Ni and Cu produced structures which were not as well graphitized.[3,4] Metal mixtures (like Fe-Mo, Fe-Co, Co-Ni, and Co-Mo) have also been used for nanotube synthesis.[5] However, the mechanism for bi-metallic catalysts affect the carbon nanotube growth is still poorly understood.

In this work, we investigated multi-wall carbon nanotube (MWNT) growth on $CaCO_3$ supported T(T=Fe, Co, or Ni)-Mo catalyst with different hydrocarbon source. The choice of the supporting material has been found to be critical for the scalable cCVD as well. The utilization of $CaCO_3$ as support has many advantages over conventional supports like SiO_2, Al_2O_3. The removal of the support and the catalysts of the catalytic system T-Mo/$CaCO_3$, can be easily done by mild acid treatment, washing with distilled water and filtration, which greatly reduces the purification cost. Additionally multi-step processes and hazardous chemicals are avoided. The effect of the reaction conditions on the characteristics of the nanotubes (diameter, crystallinity) was also investigated.

2. EXPERIMENTAL DESIGN

The T(T=Fe, Co, or Ni)-Mo/CaCO$_3$ catalyst was prepared as described elsewhere.[6] The T/Mo molar ratio was kept at 1:1 and the total metal loading is 2 or 5 wt%. First, the weighted amount of (NH$_4$)$_6$Mo$_7$O$_{24}$· 6H$_2$O were dissolved into ethanol under agitation and CaCO$_3$ was added to the solution after the metal salts were completely dissolved. Then the metal salts Fe(NO$_3$)$_3$ · 9H$_2$O, Co(CH$_3$COO)$_2$ · 4H$_2$O or Ni(NO$_3$)$_2$· 6H$_2$O were added into the mixture, respectively. The catalyst was further dried at about 130 °C overnight, and further calcined at 500 °C to decompose the nitrates from the catalyst.

Carbon nanotubes were synthesized on the T-Mo/CaCO$_3$ catalysts with cCVD approach using hydrocarbon source.[7] Around 200 mg of the catalyst was uniformly spread on a graphite susceptor and placed in a quartz tube positioned horizontally inside a radio frequency (RF) inductive furnace. After purging the system with N$_2$ (200 mL/min) for 10 min, RF heating at 350 kHz was applied to the graphite susceptor that contains the catalyst.[8] Hydrocarbon gas was introduced (methane at 30 mL/min, ethylene at 15 mL/min, acetylene at 3.3 mL/min) for 30 minutes when the set temperature reached. The as-produced CNTs were purified in one single step using diluted HCl solution with sonication. The CNTs obtained at different conditions were characterized by using transmission electron microscopy (TEM), thermogravimetric analysis (TGA), and Raman scattering spectroscopy.

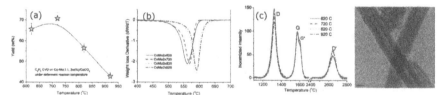

Figure 1. a) The yield of carbon nanotube grown from C$_2$H$_2$ pyrolysis on Co-Mo(1:1, 2wt%)/CaCaO$_3$ catalysts as function of reaction temperature; b) the derivative TGA curves of the CNTs obtained at different reaction temperature; c) Raman scattering spectra for the tubes synthesized at different temperature; d) the TEM images of the tubes obtained 720 °C.

3. RESULTS AND DISCUSSION

3.1 Co-Mo(1:1, 2wt%)/CaCO$_3$ under deferment reaction temperature

First, we investigated the influence of the reaction temperature on the growth of CNTs from the Co-Mo(1:1, 2wt%)/CaCO$_3$ catalyst using C$_2$H$_2$ as carbon source. The CNT yield was displayed as a function of temperature in Figure 1a. The optimal reaction temperature for the carbon nanotube growth on this catalyst is around 720 °C. The reaction temperature should be keep <750 °C as synthesis carried out above 800 °C under similar reaction conditions led to the formation of amorphous soot and carbon nanoparticles mixed with nanotubes which would prohibit the further decomposition of the carbon source on the catalytic particles and stop the CNT growth.[9]

The crystallinity of the carbon nanotube products can be evaluated by their combustion temperature (T$_0$), determined from the first order derivative of the weight loss curve. Usually small diameter CNTs or small bundles have lower combustion temperatures. Also the combustion temperature is an indicator of the crystallinity and the presence or lack of defects in the walls of the nanotubes. The CNT produced higher temperature has higher crystalline, as indicated by the higher T_0 (Figure 1b). The full width at half maximum of the combustion valley also decreases as increasing the reaction temperature.

The resonance Stokes Raman spectra from the CNT produced at different temperatures were displayed in Figure 1c. The characteristic bands for MWNTs are the D band, the G band and the D' band. The origin of the G band is a first-order Raman scattering while the D and D' bands have their origin from the double resonance.[10] The D band is associated with defects and impurities in the carbon nanotubes. The G band corresponds to the stretching mode of the C-C bond in the graphite plane. By deconvoluting the G peak (Figure 4b) it can be observed the presence of another band, G*, at about 1605 –1610 cm^{-1}, observed in all the nanotube samples. The G* band has its origins in the disorders induced in the crystalline structure and the finite size of the crystalline domains.[11] The last important mode observed in the Raman spectrum of MWNTs is the D' band. This mode is a second-order harmonic of the D band, which is often present between 2450 and 2650 cm^{-1}. The D' band is highly dispersive and usually associated with the degree of crystallinity of the carbon nanotubes. Generally, the ratio between the intensity of the G and the D bands (I_G/I_D) can be used to show the presence of structural defects and carbonaceous products with non-crystalline properties.[12] The D band intensity deceases with the increase in the reaction temperature, indicating that higher reaction temperature will help improve the crystallinity of CNTs, that is to say, the defects on the walls can be fixed at high annealing temperature. Additionally, the D mode is larger than the G mode. These features are typical for amorphous carbon deposits.

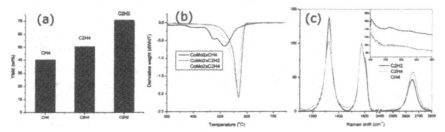

Figure 2. The CNT yield (a), the derivative TGA pattern (b), and the Raman spectra (c) of the tube obtained with different carbon source.

3.2 Different carbon source: C_2H_2, C_2H_4, and CH_4

Different carbon precursors have distinct decomposition rate and chemical composition, which might affect the growth of CNTs. Narrowly (n,m) distributed SWNTs can only be obtained under high-pressure CO or vacuumed C_2H_5OH and CH_3OH.[13] The carbon precursor chemistry may also play an important role to obtain narrowly (n,m) distributed SWNTs. Here we chose CH_4, C_2H_4 and C_2H_2 as carbon precursors. As seen from Figure 2a, the CNT yield increases from 40 wt% with CH_4 to 70 wt% with C_2H_2. Unsaturated hydrocarbons such as C_2H_2 usually have much higher yields and higher deposition rate than more saturated gases (e.g. 100 times that of C_2H_4).[14] Saturated carbon gases tended to produce highly graphitized filaments with fewer walls compared with unsaturated gases. Thus, hydrocarbons such as CH_4 and CO are commonly used for SWNT growth.[15,16]

The TGA results (Figure 2b) show that the CNT produced with CH_4 has two major decomposition peaks at 460 °C and 520 °C, respectively. It indicates the CNTs synthesized with CH_4 can be classified into two groups with different thermal stability. The lower T_0 might correspond to SWNT bundles, as suggested by the Raman result (inset of Figure 2c). The CNTs grown with C_2H_4 (561 °C) and C_2H_2 (569 °C) have pretty close combustion temperatures, but the C_2H_4 produced CNTs have a wider combustion temperature range, suggesting using C_2H_2 can produce higher wall-number/crystallinity CNTs on Co-Mo/CaCO$_3$ catalysts.

Interestingly the G/D ratio of the CNT produced from C_2H_2 is also close to that grown with C_2H_4. The CNTs synthesized from CH_4 has relatively higher G/D ratio (Figure 2c), consistent with the TGA results which show a pretty broad combustion temperature range. The G' intensity of the CNTs decreases in the order $CH_4 > C_2H_4 > C_2H_2$.

3.3 Different metal loading of Co-Mo(1:1)/CaCO$_3$ 2wt% *versus* 5wt%

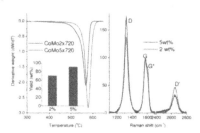

Figure 3. (a) The derivative TGA profiles of CNTs obtained on the catalysts with different metal loading, the inset shows the yield; (b) The Raman spectra of the tubes grown on the catalysts with different metal loading.

The dispersion of metal particle on the support is strongly dependent on the interaction between them as well as the specific surface area. The BET surface area measurement with krypton adsorption for the support CaCO$_3$ powder was found to be only 43.2 m^2/g. As seen from the inset of Figure 3a, the yield was only increased by 28% when the metal loading increases from 2wt% to 5wt% (150%). So, for the chosen support with certain surface area, further increasing metal catalyst loading (> 5wt%) is not an efficient way to improve the CNT yield.

The TGA measurement revealed that the CNTs grown on the 5wt% catalyst have higher combustion temperature than those produced from the 2wt% catalyst. It can be understood from the fact that with the higher metal loading the metal particles tend to aggregate and form large size clusters, as a consequence, the large diameter tubes, which have higher combustion temperature, can grow out from them. The Raman scattering spectra display a higher G/D ration from the tubes obtained from the 2wt% catalyst as compared with those from the 5wt% catalyst.

3.4 Fe-Mo, Co-Mo, and Ni-Mo

The metallic nanoparticles like Fe, Co, Ni are commonly used for the CNT synthesis due to their high activity as well as high selectivity. Usually the second metal is added to improve the CNT yield. [17] With C_2H_2 as carbon source, the Co-Mo/CaCaO$_3$ catalyst (70 wt%) shows relatively higher CNT yield as compared with the CaCO$_3$ supported Fe-Mo (65 wt%) and Ni-Mo (61 wt%) catalysts (inset of Figure 4a). The performance of catalyst system in the CNT growth strongly depends on many reaction parameters like the type of carbon source, reaction temperature and flow rate. By comparing Ni, Co, and Fe/Mo as the catalyst material on an Al support layer, Seidel *et al.* found that Ni-Mo allows the synthesis of SWNTs from methane CVD in the lowest temperature range.[18] The difference of the growth behaviors of Fe, Co, and Ni may be related to their ability to form catalyst particles with the right size and shape. The shape of the catalyst particles is largely affected by their wetting behavior on a specific substrate. It has been reported that hydrogen leads to much better wetting.[19] A strong interaction between catalyst atoms and the catalyst support or even a slight inter-diffusion will hinder the formation of appropriately shaped catalyst particles. The concept of the importance of the catalyst support on SWNT growth becomes more obvious if one regards the limited choice of substrates (*e.g.*, SiO_2, Al_2O_3, MgO) on which SWNTs and also high quality MWNTs can be grown.

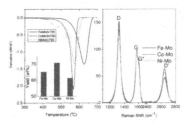

Figure 4. The effluence of the metal composition on the carbon yield (inset in a) and their thermal stability (a) as well as the Raman scattering profiles (b).

The thermal stabilities of the CNTs grown on different catalyst were show in Figure 4a. Interestingly the CNTs produced on Ni-Mo/CaCaO₃ catalyst have two combustion temperatures, indicating two types of structure with different thermal stability. The tubes synthesized on the Fe-Mo catalyst have higher thermal stability than those obtained on the Co-Mo system. The Raman spectra in Figure 4b revealed that intensities of G and D' bands decrease in the order Fe> Ni>Co.

In our T-Mo/CaCO₃ catalytic system, the element Mo probably might play two different roles in the CNT synthesis. First, Mo can act as a secondary support for the active metal like Fe here. The active metal particles docked on the relatively large Mo particles which were supported by the MgO powders.[20] Second, Mo can also participate in the breakup of the hydrocarbon molecules. It has been reported that the use of conditioning catalyst Mo/Al₂O₃ caused more double wall carbon nanotubes than SWNTs to grow,[21] possibly because of an increase in the amount of the active carbon species.[22] Carbon-containing radicals are much more active for the carbon tube formation, since they greatly reduce the reaction potential.

ACKNOWLEDGEMENT
    This work was financially supported by the U.S. Department of Energy (Grant No. DE-FG 36-06 GO 86072).

REFERENCES

[1] J.A. Kong, A.M. Cassell, H. Dai, *Chem. Phys. Lett.* 292, 567 (**1998**).
[2] P. Nikolaev, M.J. Bronikowski, R.K. Bradley, F. Rohmund, D.T. Colbert, K.A. Smith, R.E. Smalley, *Chem. Phys. Lett.* 313, 91 (**1999**).

[3] A. Kukovecz, Z. Konya, N. Nagaraju, I. Wiullems, A.Tamasi, A. Fonseca, J. B. Nagy, I. Kiricsi, *Phys. Chem. Chem. Phys.* 2, 3071 (**2000**).

[4] V. Ivanov, J.B. Nagy, P. Lambin, A. Lucas, X. B. Zhang, X. F. Zhang, D. Bernaerts, G. antendeloo, S. melinckx, J. anlanduyt, *Chem. Phys. Lett.* 223, 329 (**2002**).

[5] J. Kong, H.T. Soh., A. M. Cassell, C. F. Quate, H.J. Dai, *Nature* 395, 878 (**1998**).

[6] E. Couteau, K. Hernardi, J.W. Seo, L. Thiên-Nga, Cs. Mikó, R. Gaál, *et al*, *Chem. Phys. Lett.*, 378, 9-17, (2003).

[7] T. Schmitt, A. S. Biris, D. Miller, A. R. Biris, D. Lupu, S. Trigwell, Z. U. Rahman, *Carbon*, 44(10), 2032-2038 (2006).

[8] A. R. Biris, A. S. Biris, D. Lupu, S. Trigwell, E. Dervishi, Z. Rahman, P. Marginean, *Chem. Phys. Lett.* 429(1-3), 204-208 (2006).

[9] B. Louis, G. Gulino, R. Vieira, J. Amadou, T. Dintzer, S. Galvagno, G. Centi, M.J. Ledoux, C. Pham-Huu, *Catalysis Today* 102–103, 23–28 (2005).

[10] E.F. Antunes, A.O. Lobo, E.J. Corat, V.J. Trava-Airoldi, A.A. Martin and C. Veríssimo *Carbon* 44, 2202–2211(2006).

[11] R. A. Afre, T. Soga, T. Jimbo, Mukul Kumar, Y. Ando and M. Sharon *Chemical Physics Letters* 414, 6-10 (2005).

[12] Y.D. Lee, H.J. Lee, J.H. Han, J.E. Yoo, Y-H. Lee, J.K. Kim, S. Nahm, B-K Ju., *J. Phys. Chem. B* 110, 5310-5314 (2006).

[13] B. Wang, C. H. Patrick Poa, L. Wei, L.-J. Li, Y. Yang, Y. Chen *J. Am. Chem. Soc.* 129, 9014-9019 (2007).

[14] R. T. Baker and P. Harris, in *"Chemistry and physics of carbon"* (J. P. L. Walker and P. A. Thrower, Eds.), Vol. 14, p. 83. Dekker, New York/Basel, 1978.

[15] Kong, J.A.; Cassell, A.M.; Dai, H. *Chem. Phys. Lett.* 1998, 292, 567.

[16] W. E. Alvarez, F. Pompeo, J. E. Herrera, L. Balzano, D. E. Resasco, *Chem. Material.* 14, 1853(2001).

[17] E. Couteau, K. Hernadi, J. W. Seo, L. Thiên-Nga, Cs. Mikó, R. Gaál, and L. Forró, *Chemical Physics Letters* 378, 9–17 (2003).

[18] R. Seidel, G. S. Duesberg, E. Unger, A. P. Graham, M. Liebau, F. Kreupl *J. Phys. Chem. B* 108, 1888-1893(2004).

[19] H. Kanzow, A. Ding, *Phys. ReV. B* 60, 11180 (1999).

[20] D.E. Resasco, W.E. Alvarez, F. Pompeo, L. Balzano, J.E. Herrera, B. Kitiyanan, A. J. Borgna, *Nanoparticle Res.* 4, 131-136 (**2002**).

[21] N. R. Franklin, H. Dai, *Adv. Mater.* 12, 890-894 (**2000**).

[22] M. Endo, H. Muramatsu, T. Hayashi, Y. A. Kim, M. Terrones, M. S. Dresselhaus, *Nature* 433, 476 (**2005**).

Mater. Res. Soc. Symp. Proc. Vol. 1142 © 2009 Materials Research Society   1142-JJ05-10

Spectroscopic Characteristics of Differently Produced Single-Walled Carbon Nanotubes

Zhongrui Li,[*] Yang Xu, Enkeleda Dervishi, Viney Saini, Meena Mahmood, Olumide Dereck Oshin, Alexandru S. Biris[*]

Nanotechnology Center and Applied Science, University of Arkansas at Little Rock, Arkansas, 72204

ABSTRACT

Four differently produced single-walled carbon nanotube (SWNT) materials (by arc discharge, HiPco, laser ablation, and CoMoCat method) with diameters ranging from 0.7 to 2.8 nm were investigated by using multiple characterization techniques, including Raman scattering, optical absorption, X-ray absorption near edge structure, along with X-ray photoemission. The vibrational spectroscopies revealed that the diameter distribution and the compositions of metallic and semiconducting tubes of the SWNT materials are strongly affected by the synthesis methods. Similar sp^2 hybridization of carbon in the SWNT structure oxygenated can be found but different surface functionalities are introduced while the tubes are processed. All SWNTs exhibited stronger plasmon resonance excitations, lower electron binding energy compared to graphite. These SWNT materials also exhibit different valence band X-ray photoemission features which are affected by the nanotube diameter distribution and metallic/semiconducting composition.

1. INTRODUCTION

Single-walled carbon nanotubes (SWNTs) are usually produced by several different methods, *i.e.* arc discharge,[1] laser ablation,[2] chemical vapor deposition (like HiPco, CoMoCat).[3,4] All currently available synthesis methods result in mixtures of semiconducting and metallic nanotubes, typically bound together in bundles, and in major impurities like amorphous carbon and metallic catalyst particles. So differently produced SWNTs usually possess diverse electronic properties. A first step in a systematic approach towards improved SWNT production, application and tailoring of their electronic and mechanical properties is a feedback of information coming from reliable characterization techniques.

Thus, the use of more techniques is required to obtain a more comprehensive picture of SWNTs. In this work, we utilized Raman scattering, optical absorption, X-ray absorption near edge structure, and X-ray photoemission at both valence band and C 1s shell to systematically determine the electron properties of single-walled carbon nanotubes synthesized with different methods, including electric arc discharge, HiPco, laser ablation, and CoMoCat. The nanotube diameter distribution and the composition of metallic and semiconducting tubes play an important role in explaining the experimental results.

2. EXPERIMENTAL

Four different SWNT samples were studied: (a) SWNTs generated with arc discharge from Carbon Solutions Incorporation;[5] (b) laser ablation SWNTs tubes@Rice;[6] (c) HiPco-SWNTs from Carbon Nanotechnologies Incorporation,[7] and (d) CoMoCat-SWNTs made using a silica-supported Co and Mo catalyst.[8] The Raman spectra were obtained in a Jovin Yvon-Horiba Lab Raman system. The 1.96 eV irradiation of a He-Ne laser was used as the excitation source. The optical absorption was performed on a Shimadzu double beam spectrometer UV-3600. C K-edge XANES measurements were carried out with the total electron yield mode under UHV condition (~10^{-10} torr). The calibration of photon energy is performed with the carbon K-edge π^* transition of graphite reference sample, located at 285.35 eV. These spectra have been processed through standard pre- and post-edge normalization routines. XPS data were achieved using monochromatic Al K_α excitation (1486.6 eV) under ~2.0×10^{-9} Torr. Quantification of the surface composition was carried out by integrating the peaks corresponding to each element with aid of the Shirley background subtraction algorithm, and then converting these peak areas to atomic composition by using the sensitivity factors provided for the each element.

3. RESULTS AND DISCUSSION

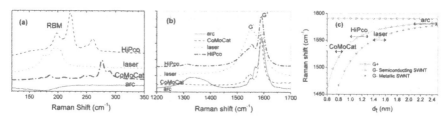

Figure 1. Raman scattering spectra (RBM band (left), D and G bands (middle)) from the single-walled carbon nanotubes produced with different method including arc discharge, HiPco, laser ablation and CoMoCat. A He-Ne laser (1.96 eV) was used as excitation source. (right) Diameter distributions of differently produced single-walled carbon nanotubes, and tube diameter dependence of ω_{G+} and ω_{G-} for semiconducting and metallic single-walled carbon nanotubes.

(1) Raman Scattering

Since the observable radial breathing mode (RBM) comes predominantly from tubes in resonance with laser energy, Figure 1(left) specifies the nanotubes observable for a 1.96 eV laser irradiation. There is a linear relation between RBM frequency and inverse diameter $\nu_{RBM} = C_1/d + C_2$, where C_1(234 cm^{-1}) describes the RBM frequency of an individual free-standing nanotube and C_2 (13 cm^{-1}) takes in to account the interaction with the local environment.[9] Obviously, the HiPco and laser ablation methods can produce larger diameter SWNTs with a wide diameter distribution (HiPco 1.0~1.4 nm, laser ablation 1.3~1.6 nm, respectively), while the CoMoCat SWNT sample dominates two types of semiconducting nanotubes (6, 5) and (7, 5), with diameter around 0.8 nm.[10] However, the arc discharge SWNT sample does not show any remarkable RBM features, likely because the laser energy is too far away from the Raman resonance of the arc discharge SWNTs.

A highly dispersive disorder-induced D band appears in the Raman spectra of graphite-like materials through a double resonance process.[11] The D band intensities of the CoMoCat, laser ablation and HiPco samples are tiny, suggesting high quality (Figure 1(middle)). The broad high D band feature of the arc discharge SWNT sample might originate from amorphous carbon or more defects in tubes, but also due to a bad resonance condition, as also seen in the poor RBM feature.

Unlike graphite, the tangential G mode in SWNTs gives rise to a multi-peak feature. The lower frequency G$^-$ band feature for metallic SWNTs is observed to have a Breit-Wigner-Fano (BWF) lineshape (that accounts for the coupling of a discrete phonon with a continuum related to conduction electrons), though the G$^-$ feature for semiconducting tubes remains Lorentzian. As seen in Figure 1(middle), the laser ablation sample exhibits a significant broadening asymmetric G$^-$ peak with a BWF lineshape while the broadening of G$^+$ is minor. This is typical metallic character of many graphite materials. The splitting between the frequency of the upper and lower G-band peaks provides a determination of the tube diameter d_t through the relation $\omega_{G+}-\omega_{G-}=C/d_t$, where C=45.8 cm^{-1}nm^2 for semiconducting tubes and 79.5 cm^{-1}nm^2 for metallic nanotubes.[12] The frequency differences between G$^+$ and G$^-$ increase in the order: arc discharge (26 cm^{-1}) < HiPco (34 cm^{-1}) < laser (41 cm^{-1}) < CoMoCat (56 cm^{-1}). So based upon the diameter dependence of the G$^+$ and G$^-$ frequencies for semiconducting and metallic tubes (Figure 1(right)),[13] we can expect that most of the CoMoCat and arc discharge tubes are semiconducting while the laser ablation bulk sample should contain more metallic tubes.

44

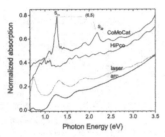 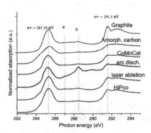

Figure 2. (left) Optical absorption of single-walled carbon nanotubes produced with arc discharge, Hipco, laser ablation and CoMoCat method after 2 hour sonication with NaDDBS (1 g/L) and 1 hour centrifugation at 27 000 ×g. (right) C 1s X-ray absorption near edge structure spectra of C reference compounds and single-walled carbon nanotubes materials produced with different approaches. See text for the description of the vertical lines.

(2) Optical Absorption

Figure 2(left) shows the optical absorption spectra of the individual tube suspensions. A correction factor should be used due to excitonic effects when determining the mean diameter from the S_{11} peak. [14] The following equation was used to determine the mean diameter from the S_{11} peak: $E_{S_{11}} - X = 2\gamma R / d$, where X (~0.154 eV) is the correction factor, R (1.407 Å) is the C-C bond length, and γ (~3 eV) is the tight binding nearest neighbor overlap integral. The assignment of the CoMoCat and HiPco tubes are very similar to the ones already published by Bachilo et al. [15] The CoMoCat sample shows two dominant structures: (6, 5) and (7, 5). Together, these account for 75% of the semiconducting tubes. No metallic tubes were observable in the corresponding scan area. The HiPco SWNT material has a wide diameter distribution, covering metallic tubes (9, 4), (12, 6) and (13, 4), and semiconducting tubes (11, 3), (10, 3), (8, 7), (8, 6), (7, 6) and (6, 5). For the laser ablation SWNT material, we expect to excite more metallic tubes according to *Kataura* plot because they have diameters ranging from 1.0 to 1.6 nm. It contains metallic tubes, (17, 2), (18, 0) and (12, 9). The other assignment is very similar to the one which was already obtained by Lebedkin et al. [16] The mean diameter of the arc discharge SWNTs (around 2.3 nm) is even larger than that of the laser ablation material, as suggested by the fact that the arc discharge SWNTs have smaller energy gaps between their E_{11} and E_{22}.

(3) X-ray Absorption Near Edge Structure

XANES measures the core hole-perturbed the density of states (DOS) within a few atomic sites of the excited atom. [17] The XANES spectra of the SWNTs samples obtained by different synthesis procedures are also shown in Figure 2(right). The peaks around 285.35 eV and 291.5 eV can be assigned to the excitation of the core electron into the π^* and σ^* states in the conduction band, due to the sp^2 hybridization of carbon in the graphite structure. Lines *a* and *b* are attributed to transitions from C 1s to σ^* states in C-H bonds and σ^* states in C-C localized defects respectively [18] or can be assigned to oxygenated surface functionalities introduced while the tubes were processed. These correspond to σ^* C=O and σ^* C-O resonances. [19] The feature *a* does not derive by the inherent graphite electronic structure, but is rather due to the presence of surface states, such as those produced by adsorption of residual oxygen. [20]

Clearly, the entire set of SWNT samples does not show the broad peak feature of the amorphous carbon, evidencing good defined structure in all of them. Nevertheless the HiPco sample exhibit the more intense feature in the structure at the region between 297-289 eV, probably associated with the poor cleanness procedure to obtain the corresponding SWNT. Similar to HOPG, the peak *b* vanishes in the rest of the carbon nanotube samples (*i.e.* CoMoCat, laser ablation and arc discharge). Finally, the main features associated with π* and σ* states are similar on the same cited NT samples indicating the similar sp² hybridization of carbon in the SWNT structure.

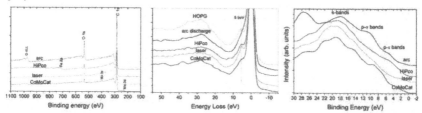

Figure 3. (left) The full survey of X-ray photoemission spectra of single-walled carbon nanotubes synthesized with different method. The structures around 745 and 1000 eV are due to oxygen and carbon Auger transitions, respectively. (middle) The C 1s core energy loss spectra for the single-walled carbon nanotubes synthesized with arc discharge, Hipco, laser ablation and CoMoCat approaches. The spectra have been normalized to the C 1s main peak and relocated with the loss energy of the main peaks all being zero. (right) The valence band X-ray photoemission spectra of single-walled carbon nanotubes synthesized with different methods.

(4)   Core Level XPS

The full energy surveys of X-ray photoemission spectra are shown in Figure 3(left) for the differently synthesized SWNTs. The structures around 745 and 1000 eV are due to oxygen and carbon Auger transitions, respectively. The C 1s peak shifts to lower about 0.2 eV for the carbon nanotubes (arc discharge 284.7 eV, HiPco 284.5 eV, laser ablation 284.7 eV, and CoMoCat 284.5 eV), as compared with that of HOPG (284.8 eV). The negative shift of the C 1s peak may be ascribed to weaker C-C binding caused by the curvature of graphene sheets. Additionally, structural information is given via chemical shift due to atomic neighbor distributions, including variations due to *e.g.* disorder and defects. A distribution of different tube diameters and contaminants must also be considered.

The C 1s line of carbon nanotubes is typically broad and asymmetric (Figure 3(middle)). A core level lineshape can be due to the following contributions: core-hole lifetime, excitation of vibrations and electron-hole pairs, and surface core level shift. The C 1s peak widths (FWHM) of the arc discharge, HiPco, laser ablation and CoMoCat SWNT samples are 0.88, 0.81, 0.86, and 0.91 eV, respectively, considerably wider than that of HOPG (about 0.56 eV). The larger FWHM is regularly observed for the nanotubes of various diameters, and may indicate a shorter lifetime of the holes of C 1s photoemission in comparison to that of HOPG. This has been attributed to the metallicity of the samples,[21] to an exotically enhanced lifetime broadening,[22] and to oxidation.[23] For cases of semiconducting tubes like CoMoCat, small charging effects should also be considered, which may not be simple to discern from the Fermi energy region spectra due to the slowly varying shape of the valence density of states.[24]

The normalized C 1s energy loss spectra for the carbon nanotubes and HOPG demonstrate a sharp peak about 5.9 eV above C 1s peak, which corresponds to π excitation. A wide loss feature around 27 eV can be attributed to collective σ+π plasma. The relative size of the loss peaks seems to depend upon several factors – the primary one being the number and character of DOS at or near the Fermi edge of the system.[25] Close study of loss size and type suggests that the best source of a system that provides a plasmon-dominated loss spectrum of relatively large size is a spherically symmetric (*i.e.* s state) or at most, p-type valence band density at or near the Fermi edge that is relatively free from any localized, directed, valence band density (*e.g.*, d-type) coming from the low binding energy parts of the valence region.

(5) Valence Band XPS

Figure 3(right) displays the valence band photoemission spectra of the SWNT materials close to the Fermi level. Note that surface states may well be a major contributor to the observed valence-band spectra. Surface atomic ratio analysis implies that less than 4% of atoms sampled are oxygen (except the arc discharge SWNTs sample with surface oxygen content of 12%). If the ratio of oxygen to carbon cross section for valence states is between 1 and 2, then from 4% to 2% of the observed valence-band structure is due to surface oxygen atoms. Thus the contribution of oxygen in valence band photoemission is very small. For the arc discharge SWNT sample with about 12% surface oxygen content, likely the oxygen contamination entirely accounts for the peak centered near 27 eV in the valence band spectrum (Figure 3(right)) due to O 2s excitation. On the other hand, the O 2p structure is significantly attenuated with respect to the O 2s feature due to its smaller cross section.[26] The broad trapezoid structure around 19 eV is the C 2s band.

The binding energy region between 2.0 and 9.0 eV is contributed by the graphene 2p-π electrons, which is overlaps with the top of 2p-σ electrons. The differently synthesized SWNT materials demonstrate quite different valence band photoemission structures. The difference might be explained due to their different diameters and metallic/semiconducting tube composition. For instance, the intensity in the binding energy region between 2.0 and 7.6 eV is significantly lower for the CoMoCat SWNT sample than the HiPco SWNT material. This can be understood with the fact that the CoMoCat SWNTs with small diameters have larger curvature, and then result in low 2p-π electron density. The 2p electrons of metallic nanotubes are highly mobilized, so no significant valence band structure between 2.0 and 14 eV was seen in the laser ablation SWNT material. This weak chirality dependence of photoemission cross section is also in well consistent with the theoretical prediction for metallic SWNTs.[27]

The energy gap of carbon nanotubes may be almost zero and some densities of states appear near the Fermi level, depending on the size of carbon nanotubes.[28] All the SWNT materials in this study show some densities of states near the Fermi level at room temperature, as seen in the inset of Figure 3(right). The valence band edge shifts to the higher binding-energy side in the sequence: laser ablation < HiPco < CoMoCat, implying the increase of band gap. This sequence is exactly consistent with the decrease in the electric conductivity of these SWNT materials. This is further evidence of higher electron density of metallic SWNT near Fermi edge.

4. CONCLUSIONS

As the conventional tools for SWNT characterization, Raman scattering suffers from the laser energy dependence limitation, and optical absorption has poor resolution for individual tubes, especially insensitive to the chirality. X-ray analysis can provide elemental resolved insight of physical and chemical properties. Combining the Raman scattering and XANES characterizations, we found that the CoMoCat and laser ablation tubes are of high quality; CoMoCat tubes show few RBM and optical absorption peaks, the collective $\sigma+\pi$ plasma excitation XPS peak above C 1s is very broad, which indicates more semiconducting tubes with narrow diameter distribution. Laser ablation tubes exhibit pronounced BWF peak and a sharp π excitation peak about 5.9 eV above C 1s peak, which suggests more metallic tubes. Arc discharge and HiPco tubes usually have larger diameter. Arc discharge tubes display strong D-band feature and C-O X-ray absorption around 288.6 eV, which are associated with more defects. HiPco tubes have many optical absorption peaks and the more intense feature at the region between 297-289 eV, also suggesting a very broad diameter distribution of semiconducting tubes. The overall shape of the valence band emission and core energy loss spectra of SWNTs show the characteristic features of carbon atoms in sp^2 hybridization, as in HOPG. The metallic and semiconducting carbon nanotubes also exhibit different photoemission properties near Fermi level. Differently synthesized SWNTs possess different physical and chemical properties, accordingly leading to their applications in different field. CoMoCat nanotubes might find wide applications in bio-sensor, novel transistor; HiPco SWNTs, bearing a wide range of direct bandgaps matching the solar spectrum and strong photoabsorption from infrared to ultraviolet, could be good media for photovoltaic cell. The laser ablation SWNTs can be excellent material for electric conducting wire in nano-devices or semitransparent conductor.

REFERENCES

[1] Ebbesen, T. W.; Ajayan, P. M. *Nature* **1992**, *358* (6383), 220-222.
[2] Guo, T.; Nikolaev, P.; Thess, A.; Colbert, D. T.; Smalley, R. E. *Chem. Phys. Lett.* **1995**, *243* (1, 2), 49-54.
[3] Kitiyanan, B.; Alvarez, W. E.; Harwell, J. H.; Resasco, D. E. *Chem. Phys. Lett.* **2000**, *317* (3, 4, 5), 497-503.
[4] Endo, M.; Takeuchi, K.; Igarashi, S.; Kobori, K.; Shiraishi, M.; Kroto, H. W. *J. Phys. Chem. Solids* **1993**, *54* (12), 1841-1848.
[5] Itkis, M. E.; Perea, D. E.; Niyogi, S.; Love, J.; Tang, J.; Yu, A.; Kang, C.; Jung, R.; Haddon, R. C. J. f *Phys. Chem. B*, **2004**, *108*(34), 12770-12775.
[6] Lebedkin, S.; Schweiss, P.; Renker, B.; Malik, S.; Hennrich, F.;Neumaier, M.; Stoermer, C.; Kappes, M. M., *Carbon*, **2002**, *40*, 417-423.
[7] Nikolaev, P.; Bronikowski, M. J. R.; Bradley, K.; Rohmund, F.;Colbert, D. T.; Smith, K. A.; Smalley, R. E., *Chem. Phys. Lett.*, **1999**, *313*, 91-97.
[8] Alvarez, W. E.; Pompeo, F.; Herrera, J. E.; Balzano, L.; Resasco, D. E., *Chem. Mater.*, **2002**, *14*, 1853-1858.
[9] Kramberger, C.; Pfeiffer, R.; Kuzmany, H.; Zólyomi, V.; Kürti, J. *Phys. Rev. B*, **2003**, *68*, 235404-4.
[10] Bachilo, S. M.; Balzano, L.; Herrera, J. E.; Pompeo, F.; Resasco, D. E.; Weisman, R. B. *J. Amer. Chem. Soc.* **2003**, *125*, 11186-11187.
[11] Saito R., Jorio A., Souza Filho, A. G., Dresselhaus G., Dresselhaus M. S., and Pimenta M. A., *Phys. Rev. Lett.* **2002**, 88, 027401-4.
[12] Jorio, A.; Souza Filho, A. G.; Dresselhaus, G.; Dresselhaus, M. S.; Swan, A. K.; Ünlü, M. S.; Goldberg, B. B.; Pimenta, M. A.; Hafner, J. H.; Lieber, C. M.; Saito, R. *Phys. Rev. B* **2002**, *65*, 155412-9.
[13] Jorio, A., Pimenta, M. A., Souza Filho, A. G., Saito, R., Dresselhaus, G., and Dresselhaus, M. S., *New J. Phys.* **2003**, *5*,139.1-139.17.
[14] Liu, X.; Pichler, T.; Knupfer, M.; Golden, M. S.; Fink, J.; Kataura, H.; Achiba, Y. *Phys. Rev. B*, **2002**, *66*, 045411-8.
[15] Bachilo, S. M.; Strano, M. S.; Kittrell, C.; Hauge, R. H.; Smalley, R. E.; Weisman, R. B., *Science*, **2002**, *298*, 2361-2365.
[16] Lebedkin, S.; Arnold, K.; Hennrich, F.; Krupke, R.; Renker, B.;Kappes, M. M., *New J. Phys.*, **2003**, *5*, 140.1-140.11.
[17] Skytt, P.; Glans, P.; Mancini, D. C.; Guo, J.-H.; Wassdahl, N.; Nordgren, J.; Ma, Y. *Phys. Rev. B* **1994**, *50*, 10457-10461.
[18] Kuznetsova, A.; Popova, I.; Yates, J. T., Jr.; Bronikowski, M. J.; Huffman, C. B.; Liu, J.; Smalley, R. E.; Hwu, H. H.; Chen, J. G. *J. Am. Chem. Soc.* **2001**, *123*, 10699-10704.
[19] Banerjee, S.; Hemraj-Benny, T.; Sambasivan, S.; Fischer, D. A.; Misewich, J. A.; Wong, S. S. *J. Phys. Chem. B* **2005**, *109*, 8489-8495.
[20] Abbas, M.; Wu, Z. Y.; Zhong, J.; Ibrahim, K.; Fiori, A.; Orlanducci, S.; Sessa, V.; Terranova, M. L. Davoli, I. *Appl. Phys. Lett.* **2005**, *87*, 051923-3.
[21] Goldoni A.; Larciprete R.; Kaulich B.; Kiskinova M.; Zhang Y.; Dai H.; Sangaletti L.; Parmigiani F. *Appl. Phys. Lett.* **2002**, *80*, 2165-2167.
[22] Chen P.; Wu X.;Lin J.; Ji W.; Tan K. L., *Phys. Rev. Lett.* **1999**, *82*, 2548-2551.
[23] Ago H.; Kulger T.; Cacianlli F.; Salaneck W.R.; Shaffer M. S. P.; Windle A. H.; Friend R. H. *J. Phys. Chem. B* **1999**, *103*, 8116-8121.
[24] Bennich P.; Puglia C.; Bruhwiler P. A.; Nilsson A.; Maxwell A. J.; Sandell A.; Mårtensson N.; Rudolf P. *Phys. Rev. B* **1999**, *59*, 8292-8304.
[25] Bhom, D.; Pines, D. *Phys. Rev.* **1951**, *82*, 625-634.
[26] Yeh, J.J.; Lindau I.; *Atomic Data and Nuclear Tables* **1985**, *32*(1), 1-155.
[27] Odintsov, A. A.;Yoshioka, H., *Phys. Rev. B* **1999**, *59*, R10457-R10460.
[28] Blase, X.; Benedict, L. X.; Shirley, E. L.; Louie, S. G. *Phys. Rev. Lett.* **1994**, *72*, 1878–1881.

Mater. Res. Soc. Symp. Proc. Vol. 1142 © 2009 Materials Research Society   1142-JJ05-16

Effect of the powering frequency on the synthesis of carbon nanostructures by AC arc discharge at atmospheric pressure

Marco Vittori Antisari, Daniele Mirabile Gattia, Renzo Marazzi, Emanuela Piscopiello[1] and Amelia Montone

ENEA, Department of Advanced Physical Technologies and New Materials,
C.R. Casaccia Via Anguillarese 301, 00123, Rome, Italy
[1]C.R.Brindisi, S.S. 7 Appia Km 706, 72100, Brindisi, Italy

ABSTRACT

In this paper we report about the synthesis of single wall carbon nanohorns and highly convoluted graphite sheets by AC powered arc discharge carried out between pure graphite electrodes. The arc is ignited in air and the arched electrodes are surrounded by a cylindrical collector which collects the synthesized material and contributes to control the synthesis environment. With the purpose of studying the effect of the process variables, in this work we have explored the effect of the powering frequency on the structure of the synthesized material and on the yield of the process. Preliminary experimental results on tests carried out at constant voltage, show that the process yield is strongly influenced by the powering frequency and that higher yields are obtained at low frequency. The structure of the resulting soot has been characterized by transmission electron microscopy. Two kinds of microstructures are found by TEM observation constituted by highly convoluted graphene sheets, having locally the nanohorn morphology, and better organized nano-balls where also graphite nano-sheets can be locally found. The relative abundance of the two kinds of particles appears to depend on the powering frequency with a larger amount of the latter observed in samples synthesized at high frequency.

INTRODUCTION

Carbon nanostructures can be synthesized by the assembling of free carbon atoms in suitable experimental conditions. Arc discharge and impulsive laser ablation are the most explored experimental methods to produce carbon gas able to give rise to non-equilibrium nanostructures during the re-condensation process. In these methods carbon atoms are generated by graphite sublimation at high temperature, differently from CVD methods where the carbon supply is provided by the decomposition of a precursor molecule. Arc discharge has been used to produce for the first time fullerenes [1] and carbon nanotubes. This technique is useful to synthesize both Multi Wall Carbon Nanotubes (MWNT) [2] and Single Wall Carbon Nanotube (SWNT) [3] when electrodes constituted by a mixture of transition metal and carbon are ignited. In 1999 Iijima et al. [4] discovered another form of carbon nanostructures which can be produced starting from pure graphite: Single Wall Carbon Nanohorns (SWNH). This nanostructure is formed by single graphene curved in a "horn-like" shape with a fullerene at the tip.

Synthesis experiments are generally carried out in controlled environment provided by a closed chamber which can be evacuated and filled with a gas at the desired pressure. However it has been shown that also un-conventional environments like liquid nitrogen or water can be used to synthesize carbon nanostructures [5,6]. In previous papers we have shown that air at room pressure represents a suitable environment for nano-carbon synthesis by arc discharge [7]. The kind of carbon nanostructure resulting from the re-condensation of carbon sublimated at the arc depends on the synthesis conditions. We have been able to show that a major effect is played by the kind of arc powering and large differences have been noticed between DC and AC powering. In fact, in the former case, the main effect of the arc discharge is to sublimate carbon at the anode which condenses, in the form of nanostructured carbon at the cathode, with only a minor amount of sublimated material escaping from the arc area and condensing in the form of light soot. This behaviour has been explained considering the thermal asymmetry between anode which is hold at high temperature by direct electron bombardment and cathode which is only indirectly heated by irradiation from the anode and plasma coupling. In the case of pure graphite electrodes, transmission electron microscopy shows that the cathode deposit is rich in multi walled carbon nanotubes and the light soot is mainly constituted by bud like carbon nanohorns. In the latter case, when AC power is used to feed the arc, the anode and the cathode invert their role with the powering frequency and the above mentioned thermal asymmetry is lost. Consequently, for a large enough power applied to the arc, the electrode temperature is above the sublimation temperature for carbon and no cold points are available for carbon condensation at neither electrode. The sublimated carbon is so forced to leave the space between electrodes, and it re-condenses in the gas surrounding the arc in the form of nanometric particles. Also in this case TEM shows that these particles are constituted by carbon nanohorns. In synthesis AC powering of the electric arc between pure graphite electrodes allows to maximize the fraction of the sublimated carbon condensing in the gas surrounding the arc in the form of horned nano-particles. This synthesis method is particularly attractive, considering that the experimental apparatus is very simple, cheap, easy to use and potentially scalable to industrial level, owing to the absence of a vacuum plant. In comparison with laser ablation, the method often used to synthesize SWNH, we can notice a potentially higher productivity however coupled with a smaller degree of control. In consideration of the potential applications of SWNH in several technologies, like solvents and gases absorber [8,9], catalyst support [10,11], super-capacitors, hydrogen storage, it appears attractive to set up a better degree of control of this method. In this paper we report about the experimental results obtained by powering the electric arc with a variable frequency power supply. The effect of the powering frequency at fixed applied voltage on the process yield and on the structure of synthesized nano-carbon is reported.

EXPERIMENTAL DETAILS

The arc discharge is ignited in a specially designed experimental device able to stabilize and homogenize the electric arc between two carbon electrodes. These conditions are achieved by axially rotating at controlled angular speed the two electrodes with respect to each-other and by a constant speed advancement of one electrode toward the other. These experimental details are sufficient to ensure a stable discharge over the time of several tens of seconds and a uniform consumption of the electrodes. The system is not connected to any vacuum line and the arc can be ignited in gaseous environments at atmospheric pressure [12].
The device can be powered by different power supplies. Besides a DC power supply and a 50 Hz AC generator, we have used, in this paper, a specially designed AC power supply where the frequency can be changed in the range 32Hz-1000Hz. All power supplies are able to provide up to 50 V and 100 A.

In the experiments here described the electric arc has been ignited between cylindrical electrodes made of pure graphite (99,999 %) and having the same 6 mm diameter. The experimental set up is completed by the presence of a stainless steel collector, schematically described in figure 1.

The collector is constituted by a metallic tube having a 27.5 mm inner diameter and it is mechanically connected with the support of the lower electrode. The collector has several functions. In fact it is necessary for an efficient collection of the synthesized carbon nano-powder and influences temperature distribution and gas exchange at the arc.

During the arc, the lower electrode rotates around its axis at a speed of 30 rpm in order to have a more homogeneous discharge. The upper electrode moves down toward the lower one in order to have a continuous arc. The measurements have been performed in the voltage range of 24-28 V and in frequency range of 32-1000 Hz. All the material examined in this paper has been synthesized in air and the experiments explored the voltage range of 24-28 V, while experiments at different frequency were carried out at 32Hz, 200Hz, 600Hz and 1000Hz.

The amount of evaporated electrodes as well as the weight of the material deposited through the different condensation channels was measured after each run. The microstructure of the synthesized material was determined by transmission electron microscopy observations with a TECNAI G^2 30F TEM. The samples were prepared by dispersing the carbon nanomaterial in ethanol, followed by sonication for fifteen minutes and sprinkling onto a holey carbon grid.

DISCUSSION

Before entering into the description of the experimental results let us report about the role of the collecting cylinder on the performances of the synthesis apparatus (figure 1).

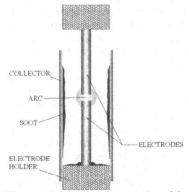

Figure 1. Schematic representation of the adopted experimental set-up

This device appears in fact of primary importance in particular in setting the thermal conditions suitable for carbon sublimation and nano-carbon synthesis. We have to report that if the device is operated without the collector, while keeping constant the other operative conditions, no carbon sublimation is observed during AC arc discharge. This is probably due to the modification of the heat dissipation channel from the hot electrode tips.

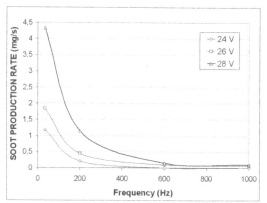

Figure 2. Soot production rate in function of the frequency of the powering current for the three samples observed.

The amount of soot collected from the collector for different values of the frequency and for fixed powering voltages is reported in figure 2. It is possible to notice how at high frequency the soot production is much smaller than at lower voltages for all the values of the applied voltage.

We want to notice that this behaviour can be correlated to a similar trend displayed by the current which results lower at high frequency. To give a first order interpretation to this behaviour we have to take into account that the current of the electric arc is depending on the cathode temperature which controls the thermal electron emission providing the charge carriers through the electrode gap. The current is so controlled by a feedback mechanism. In fact the cathode temperature, which controls the thermal electron emission, depends, among others, on the current itself. The effect of frequency is probably to interfere with this type of feedback mechanism. In fact it is reported that [13], when heat is supplied to a surface with a sinusoidal dependence on time, the surface temperature depends on the frequency and it is lower at higher frequency values. In other words, heating a surface with a power which is sinusoidal function of time, favours the heat penetration into the bulk at high frequency with a reduction of the surface temperature. Even if the situation in our experiment is more complex we can make the hypothesis that the current reduction can be correlated to a smaller surface temperature at high frequency. We want to notice that the load of the whole circuit is purely resistive so that it is not expected to change with the frequency. As far as the structure of the produced particle is concerned, results are reported in figure 3 where typical TEM images of samples synthesized at 28 V for different values of the frequency are reported. Even if a quantitative evaluation of the sample microstructure is not an easy task for this kind of sample, we want to notice that all the samples appear to be constituted by two different kinds of nano-particles. The first kind of microstructure is constituted by dahlia like nanohorned particles where nanohorns protrude from the surface, while the second kind of microstructure is more compact and often well organized graphitic structures can be observed. The former is displayed in figure 3a, while the latter can be seen in figure 3d. From the TEM observation of several samples we can conclude that the second type of particles is more abundant in the samples synthesized at higher frequency.

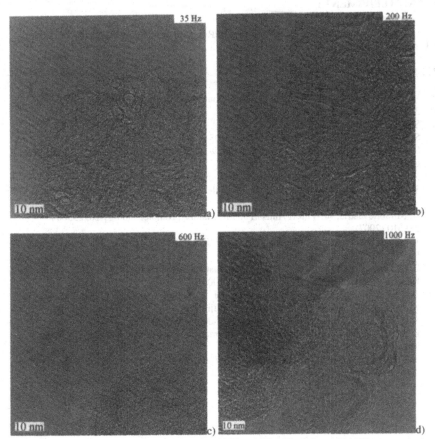

Figure 3. TEM images of the soot produced at 28 V and 35, 200, 600 and 1000 Hz respectively.

CONCLUSIONS

Arc discharge experiments carried out in air at room pressure between pure graphite electrodes powered by a variable frequency AC power supply have shown that nanohorned particles can be synthesized in the whole explored frequency range. The synthesis yield appears to depend on the powering frequency and higher yields are obtained at low frequency. Also the sample microstructure appears to be influenced by the frequency and less compact samples are obtained at low frequency. These preliminary results show that the powering frequency is a process parameter able to influence the nano-carbon synthesis by arc discharge, and further work is in progress to completely elucidate its role.

ACKNOWLEDGMENTS

Financial support by the Italian Ministry of Research through the FISR project TEPSI is gratefully acknowledged.

REFERENCES

1. W. Krätschmer, Lowell D. Lamb, K. Fostiropoulos and Donald R. Huffman, Nature 347, 354 – 358 (1990)]
2. S. Iijima, Nature 354, 56–58 (1991)
3. S. Iijima and T. Ichihashi, Nature 363, 603–605 (1993)
4. S. Iijima, M. Yudasaka, R. Yamada, S. Bandow, K. Suenaga, F. Kokai and K. Takahashi, Chem. Phys. Lett. 309, 165–70 (1999)
5. M. Vittori Antisari, R. Marazzi and R. Krsmanovic, Carbon 41, 2393–2401 (2003)
6. V. Contini, R. Mancini, R. Marazzi, D. Mirabile Gattia and M. Vittori Antisari, 87 Phil. Magazine 87, 1123–1137 (2007)
7. D. Mirabile Gattia, M. Vittori Antisari and R. Marazzi, Nanotechnology 18, 255604 - 255610 (2007)
8. J. Adelene Nisha, M. Yudasaka, S. Bandow, F. Kokai, K. Takahashi and S. Iijima, Chem. Phys. Lett. 328, 381–386 (2000)
9. E. Bekyarova, K. Murata, M. Yudasaka, D. Kasuya, S. Iijima, H. Tanaka, H. Kahoh and K. Kaneko J. Phys. Chem. B 107, 4681–4684 (2003)
10. T. Yoshitake, Y. Shimakawa, S. Kuroshima, H. Kimura, T. Ichihashi, Y. Kubo, D. Kasuya, K. Takahashi, F. Kokai, M. Yudasaka and S. Iijima, Physica B 323, 124–126 (2002)
11. S. Litster and G. McLean, J. Power Sources 130, 61–76 (2004)
12. D. Mirabile Gattia, M. Vittori Antisari, R. Marazzi, L. Pilloni, V. Contini and A. Montone, Mater. Sci. Forum 518, 23–28 (2006)
13. H.S. Carlslaw and J.C. Jaeger, *Conduction of Heat in Solids*, 2rd ed. (Oxford, UK: Oxford University Press, 1959)

Mater. Res. Soc. Symp. Proc. Vol. 1142 © 2009 Materials Research Society   1142-JJ05-31

Synthesis and Characterization of Mixed-Metal Core-Shell Nanowires

Jin-Hee Lim and John B. Wiley*
Department of Chemistry and Advanced Materials Research Institute,
University of New Orleans, New Orleans, LA 70148, U.S.A.

ABSTRACT

Template methods offer a very simple and facile approach to the controlled growth of nanowires. The templates themselves can be fabricated with a range of pore sizes and the growth of the wires within the templates can readily occur with a variety of diameters and lengths. Use of electrochemical deposition to synthesize ordered arrays of nanostructures in anodic alumina membranes (Anodic Alumina Membrane) is well studied. Electrochemical growth within the templates depends on the applied current. This allows one to not only grow nanowires within template pores but to grow nanotubes as well. If one then deposits metal within the tubes by a subsequent deposition step, core-shell nanowires can be produced. Herein, we describe the fabrication and characterization of both nanotubes and mixed-metal core-shell nanostructures produced by controlling the current density within AAM.

INTRODUCTION

The development of 1-D magnetic nanostructures has been the focus of extensive research because of their potential applications in the fields of nanodevices, optoelectronic devices, biological systems, etc [1]. For the synthesis of magnetic nanostructures, the template method is an important technique due to the simplicity of control over the diameter or length through their highly ordered uniform structures. Many kinds of materials, which have been fabricated via electrodeposition process with templates method such as semiconductor [2], polymer [3], and magnetic material [4], are investigated to study their unique properties. However, there are very few reports on the fabrication and research of nanotubes or core-shell nanostructures using a template method. Recently, metal nanotubes have been synthesized by a chemical modification of the pore of the template [5] or the method of melt-assisted template growth [6] prior to electrodeposion. Cao et al. reported on metal nanotubes (Fe, Co, Ni) fabricated by a direct electrochemical deposition technique and suggest a growth mechanism [7] for this process. Also, Wang et al. [8] and Kamalakar [9] fabricated Cu nanotubes in AAM.

In most cases a template method combined with electrodeposition is used for the fabrication of nanowires, nanotubes or metal/polymer composite nanostructures [10]. In this study, we describe the fabrication and characterization of metal nanotubes and mixed-metal core-shell nanowires in AAM using a two-step electrodeposition.

EXPERIMENT

Core-shell nanowires were grown electrochemically within porous membranes. To create the electrode used in deposition, an alumina membrane (Whatman Corp., average diameter of 200 nm) was sputtered on one side with Ag film (200-300 nm thick). Most of the surface of the Ag film was coated with glue (3M Scotch) to avoid deposition of metal; a small uncoated region of

AAM was fixed to an alligator clip. Electrodeposition was carried out at room temperature by constant current method in a two-step process on a Princeton Applied Research VMP2. A platinum wire was used as the counter electrode. Ni (Nickel sulfamate-RTU) and Au (Orotemp 24) plating solution were commercially available from Technics Inc. and Co plating solution was contained 240 g/L cobalt sulfate hepta-hydrate (99% $CoSO_4 \cdot 7H_2O$, Sigma) and 40 g/L boric acid (99.5% H_3BO_4, Alfa Aesar). Initially nanotubes were grown in one solution by deposition at –0.5 mA for 10 min, then the sample was rinsed three times with distilled water and transferred to the second solution were the core structure was grown, also at –0.5 mA over a 10 minute period. The metal nanotubes, which are fabricated by first deposition process, are fully filled by a subsequent deposition step with Au, Co, or Ni, respectively. The AAM containing metal nanostructures are etched slowly with 1.0 M NaOH solution and washed several times with distilled water resulting in well-aligned nanowires.

Field-emission scanning electron microscope (FESEM) imaging was obtained on a LEO 1530 VP. EDS was carried out using EDAX GENESIS equipped with JSM 5410 electron microscope. Powder X-ray diffraction (XRD) data were collected on a Phillips X-pert PW 3040 MPD X-ray powder diffractometer with Cu Kα radiation.

RESULTS AND DISCUSSION

Well-aligned metal nanotube and core-shell nanowire arrays are successfully fabricated by a two-step electrodeposition method as illustrated in Figure 1. The AAM templates used in this study have an average pore diameter of 200 nm. Ag film was sputtered onto the top surface of AAM; this also resulted in some silver in the pore channels. After sealing of the metal surface with glue, metal nanotube arrays, which compose the shell part of the core-shell structures, are fabricated by in the first electrodeposition step. We are able to grow both Co and Ni nanotubes by deposition at -0.5 mA for 10 min. FESEM images of metal nanotube arrays show that the synthesized nanostructures are clearly composed of well-aligned nanotubes (Figure 2). Typically, the metal nanotubes are easily aggregated and bend when the AAM is etched away by NaOH solution. These nanotubes are around 10 μm in length with a 40 nm shell thickness. The core of the core-shell nanostructures can be produced in the second electrodeposition process using

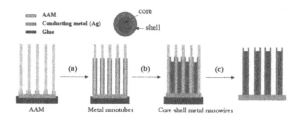

Figure 1. Procedure for preparation of metal core-shell nanowire arrays. (a):first electrodeposition, (b) second electrodeposition, (c) removing of AAM.

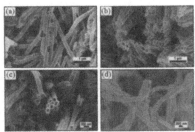

Figure 2. FESEM images of nanotubes structures. (a), (b) Co nanotubes, and (c), (d) Ni nanotube after removal of AAM.

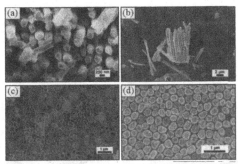

Figure 3. FESEM images of core-shell nanowires. (a), (b) NiCo, (c) CoNi, and (d) AuNi core-shell nanowires after removal of AAM.

either Co or Ni nanotubes. Figure 3 shows the core-shell nanowire arrays with an inter-wire distance of around 400 nm. The diameter of metal shell component maintains its diameter of around 40 nm and the core of the tube is completely filled. Additionally, Figure 3 (b) shows core-shell wires after they were released from the template. The length of wires (ca. 10 μm) is like that of the corresponding nanotubes. Energy-dispersive spectrometry (EDS) was used for elemental analysis of both nanotubes and core-shell nanowires after removal of AAM (Figure 4). Figure 4 (a)-(b) were measured from those nanotubes shown in Figure 2; the EDS clearly shows both Co and Ni peaks. Also, NiCo and CoNi core-shell nanowires contained both Ni and Co peaks (Figure 4 (c)-(d)). Ag peaks result from the conducting metal under the nanostructures. The FESEM image in Figure 5 (a) exhibits Au nanowires after removal of the Ni shell of a set of NiAu core-shell nanowires; the nickel was dissolved with a 1.0 M HCl solution. The EDS pattern shows that the structures are composed of Au. The ends of the wires are larger because the second electrodeposition process exceeded the length of the nanotubes. Electrochemically grown Co and Ni nanowires generally have hcp and fcc structures, respectively [11]. In our experiments, Co and Ni nanotubes do show the hcp and fcc structures, as evidenced by the XRD data in Figure 6 (b) and (c). The strong peaks in Figure 6 (a), for NiCo core-shell nanowires, are detected at $2\theta \approx 42°$ and $76°$ corresponding to the Ni (111) and (002). Also, the core-shell nanowires are showing Co peaks and intensity of peaks depends on which material is composing the core part of core-shell nanowires.

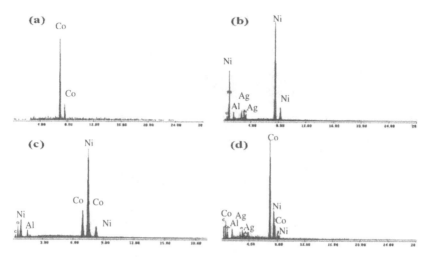

Figure 4. EDS of (a) Co nanotube, (b) Ni nanotube, (c) NiCo, and (d) CoNi core-shell nanowire after removal of the AAM.

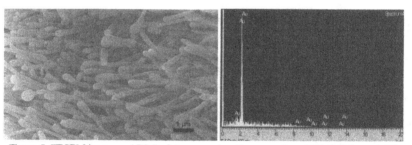

Figure 5. FESEM image and EDS of Au nanowire after removing of Ni shell from AuNi core-shell nanowire.

CONCLUSIONS

In summary, we successfully fabricated both nanotubes and core-shell nanostructures in AAM by a two-step electrodeposition method. The nanotubes were synthesized with Co and Ni and the cores were Co, Ni and Au. We can also grow core-shell nanostructures with other

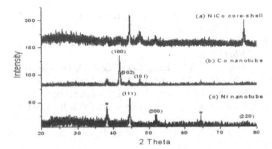

Figure 6. XRD patterns of (a) NiCo core-shell nanowires, (b) Co nanotubes, and (c) Ni nanotubes. (* indicate Ag peaks on the AAM pore bottom).

materials like nonmagnetic, soft or hard magnetic materials. It should be possible to expand this approach to the growth of core-shell structures with other materials such as semiconductors or conducting polymers for potential applications in electronics, magnetics, biological systems, or sensor materials.

ACKNOWLEDGMENTS

Support by the National Science Foundation through the NIRT program (NSF-0403673) is gratefully acknowledged.

REFERENCES

1. Timothy A. Crowley, Brian Daly, Michael A. Morris, Donats Erts, Olga Kazakova, John J. Boland, Bin Wu and Justin D. Holmes, *J. Mater. Chem.* **15**, 2408 (2005). D.H. Pack, Y.B. Lee, M.Y. Cho, B.H. Kim, S.H. Lee, Y. K. Hong, J. Joo, H.C. Hong, S. R. Lee, *Appl. Phys. Lett.* **90**, 093122 (2007).
2. Hong Jin Fan1, Woo Lee, Roland Scholz, Armin Dadgar, Alois Krost, Kornelius Nielsch and Margit Zacharias, *Nanotechnology*, **16**, 913 (2005). Liang Li, Shusheng Pan, Xincun Dou, Yonggang Zhu, Xiaohu Huang, Youwen Yang, Guanghai Li, and Lide Zhang, *J. Phys. Chem. C*, **111**, 7288 (2007). Maojun Zheng, Guanghai Li, Xinyi Zhang, Shiyong Huang, Yong Lei, and Lide Zhang, *Chem. Mater.* **13**, 3859 (2001).
3. Rui Xiao, Seug Il Cho, Ran Liu, and Sang Bok Lee, *J. Am. Chem. Soc.* **129**, 4483 (2007).
4. Zhu Liu and Peter C. Searson, *J. Phys. Chem. B*, **110**, 4318 (2006). Andrew I. Gapin, Xiang-Rong Ye, Li-Han Chen, Daehoon Hong, and Sungho Jin, *IEEE Trans. Magn.*, **43**, 2151 (2007). Samantha A. Meenach, Jared Burdick, Anurag Kunwar, and Joseph Wang, *Small* **3**, 239 (2007). Hui Pan, Binghai Liu, Jiabao Yi, Cheekok Poh, Sanhua Lim, Jun Ding, Yuanping Feng, C. H. A. Huan, and Jianyi Lin, *J. Phys. Chem. B*, **109**, 3094 (2005).
5. Woo Lee, Roland Scholz, Kornelius Nielsch, and Ulrich Gosele, *Angew. Chem. Int. Ed.* **44**, 6050 (2005). Jianchun Bao, Chenyang Tie, Zheng Xu, Quanfa Zhou, Dong Shen, and Qiang

Ma, *Adv. Mater.* **13**, 1631 (2001). Yu-Li Tai and Hsisheng Teng, *Chem. Mater.* **16**, 338 (2004).

6. Claire Barrett, Daniela Iacopino, Deirdre O'Carroll, Gianluca De Marzi, David A. Tanner, Aidan J. Quinn, and Gareth Redmond, *Chem. Mater.* **19**, 338 (2007).

7. Huaqiang Cao, Liduo Wang, Yong Qiu, Qingzhi Wu, Guozhi Wang, Lei Zhang, and Xiangwen Liu, *Chem. Phys. Chem.*, **7**, 1500 (2006).

8. Yinhai Wang, Changhui Ye, Xiaosheng Fang, and Lide Zhang, *Chem. Lett.* **33**, 166 (2004).

9. M. Venkata Kamalakar and Arup K. Raychaudhuri, *Adv. Mater.* **20**, 149 (2008).

10. Nielsch K, Castano F J, Ross C A and Krishnan R, *J. Appl. Phys.* **98**, 034318 (2005). Jinsoo Joo, Dong Hyuk Park, Mi-Yun Jeong, Yong Baek Lee, Hyun Seung Kim, Won Jun Choi, Q.-Han Park, Hyun-Jun Kim, Dae-Chul Kim, and Jeongyong Kim, *Adv. Mater.* **19**, 2824 (2007). Michal Lahav, Emily A. Weiss, Qiaobing Xu, and George M. Whitesides, *Nano Lett.* **6**, 2166 (2006). Daihua Zhang et al., *Nano Lett.* **4**, 2151 (2004).

11. Jian Qin, Josep Nogue´s, Maria Mikhaylova, Anna Roig, Juan S. Munoz, and Mamoun Muhammed, *Chem. Mater.* **17**, 1829 (2005).

Mater. Res. Soc. Symp. Proc. Vol. 1142 © 2009 Materials Research Society   1142-JJ05-37

Fabrication of Highly Porous Zinc and Zinc Oxide Nanostructures

Joshua M. LaForge[1] and Michael J. Brett[1,2]
[1]Electrical and Computer Engineering, University of Alberta, Edmonton, AB, Canada
[2]National Institute for Nanotechnology, Edmonton, AB, Canada

ABSTRACT

Glancing angle deposition (GLAD) is a physical vapor deposition technique that depends on a highly oblique flux angle to create porous, large surface area thin films via self-shadowing. Control of the deposition parameters may provide a means to tune film porosity for zinc oxide sensors and photovoltaic devices. However, the self-shadowing mechanism requires a collimated particle flux, and therefore GLAD performs best under high vacuum. Creating structured films with sputtered GLAD is difficult since the high chamber pressure (>1 mTorr) necessary to maintain the sputter plasma reduces the mean-free-path of flux particles to less than 100 mm. By using an aperture to reduce the angle subtended by the target from the perspective of the substrate, maintaining an argon plasma pressure of 1.4 mTorr, and reducing the throw distance to less than 50 mm we were able to produce structured, polycrystalline, zinc thin films via GLAD. At oblique flux angles, highly porous films consisting of randomly oriented nanorods are grown. The nanorods have diameters between 10-100 nanometers with lengths up to several micrometers. Annealing at temperatures up to 250 °C in air produces polycrystalline zinc oxide with minimal changes to the film structure. We present details of the thin film fabrication process for the convoluted nanorod film morphology. We report characterization results for films produced at several deposition angles before and after annealing using scanning electron microscopy (SEM), x-ray diffraction (XRD), and transmission electron microscopy (TEM).

INTRODUCTION

Zinc oxide (ZnO) is a suitable semiconductor material for device applications including sensing, catalysis, and photovoltaics due to its unique combination of properties: piezoelectricity, a direct band-gap of 3.37 eV, large exciton and bi-exciton binding energies of 60 meV and 15 meV respectively, optical transparency, and biocompatibility [1-3]. Fabrication of ZnO thin films with a nano/micro-structured morphology is desirable for several applications due to the increase in effective surface area. Several methods are used to successfully grow ZnO nanostructures [2, 3]: vapor phase growth, vapor-liquid-solid growth, metal organic chemical vapor deposition, hydrothermal synthesis, and RF magnetron sputtering [4].

GLAD is a physical vapor deposition technique used to control film morphology (e.g. columns, square spirals, helices, etc) and porosity by careful manipulation of the substrate orientation during deposition [5, 6]. In this study we apply GLAD to the fabrication of highly porous ZnO thin films.

Magnetron sputtering of metal-coated Si substrates at normal incidence has been used previously to fabricate ZnO nanorods [4]. Our technique forms porous Zn films with a different morphology on uncoated Si substrates. We also demonstrate preliminary SEM results that suggest that film porosity can be controlled through the GLAD deposition angle. Post-deposition thermal annealing of the Zn films leads to oxidization of Zn into ZnO with only slight changes to film morphology. This method may allow for the fabrication of ZnO films with a controlled porosity.

EXPERIMENTAL SETUP

All of the Zn films were deposited on Si (100) p-type substrates with a magnetron sputtering system using three inch diameter metallic zinc targets. The substrate holder was tilted to adjust the angle (α) between the substrate normal and target normal. Before deposition, the vacuum chamber was evacuated to a base pressure of $\sim 10^{-6}$ Torr. During deposition the chamber pressure was held between ~ 1.4 mTorr and 10 mTorr. Argon gas flowed into the system at a rate of 4.00 ± 0.04 sccm to maintain the plasma during deposition. Porous films were deposited without the addition of oxygen to the plasma.

For some films, an aperture was placed between the substrate and the sputter target to limit the angular distribution of the flux incident on the substrate. The throw distance (D) between the target and substrate was kept between 40 mm and 50 mm. A 5kW supply (Advanced Energy Pinnacle™ Plus+) was operated in pulsed DC and regular DC modes to excite the plasma. During pulsed operation, the frequency and reverse time were set to 350 kHz and 1.1 µs respectively to minimize deposition pressure.

Films were characterized using XRD (Bruker D8 Discover), TEM (JEOL JEM-2200FS) and SEM (Hitachi S4800 and JEOL 6301).

EXPERIMENTAL RESULTS

Details of the process recipes for the presented films are shown in Table 1.

Table 1: Process recipes for the presented films.

Film	α [Degrees]	D [mm]	Aperture Dia. [mm]	Press. [mTorr]	Power [W]	Voltage [V]	Pulsed
A	84	46	6	1.4	200	350	Yes
B	84	47	6	1.4	300	428	Yes
C	84	48	6	1.4	400	490	Yes
D	84	48	6	10.0	200	1157	No
E	0	40	-	1.4	200	350	Yes
F	0	44	-	10.0	200	1157	No

Instead of the expected columnar films, highly porous films of intertwined randomly oriented zinc nanorods were deposited. Several films (A, B, C) grown with different pulsed plasma powers all grew with a similar porous morphology as seen in Figure 1: . For all four films deposited through an aperture (A, B, C, D) a significant amount of void is present between randomly aligned and connected Zn nanorods. In some cases, it is clear that the sides of the nanorods are faceted (Film A). In the case of Film B, small crystallites can be seen on the surface of some of the nanorods. Within each film, nanorods of a variety of dimensions ranging from 10-100 nm in diameter and up to several micrometers in length are observed. At 10 mTorr deposition pressures and DC plasma excitation porous Zn films were also grown (Film D), but appear to be more densely packed than the films grown at lower pressures.

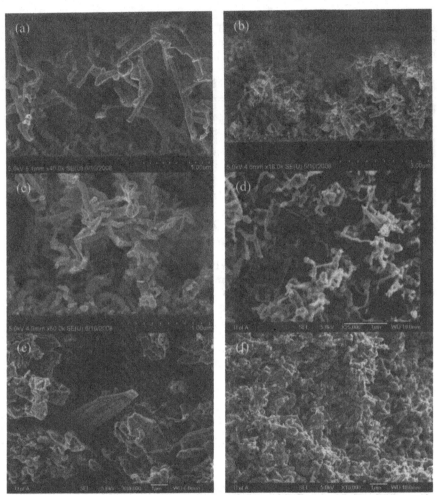

Figure 1: Films A, B, C, D, E, F from left to right, and top to bottom. SEM images of the films taken normal to the cleaved edge of the film and substrate.

Films E and F were grown during the deposition of films A and D respectively on substrates placed parallel to the target ($\alpha=0°$) at a throw distance of approximately ~40 mm. Side shots of the films taken with SEM are shown in Figure 1. Although porous, the films are much denser than the films deposited at $\alpha=84°$. The morphology of the films is different as well; the nanorods seen in the films deposited at $\alpha=84°$ have been replaced with Zn crystallites with characteristic dimensions of approximately one micrometer and have smaller aspect ratios than the nanorods in Film A-D.

63

After deposition, the films were annealed at 250 °C in air for approximately 24 hours. Comparison of the SEM images taken before and after annealing reveals that the morphology of the films remains intact. However, after annealing the surface of the nanorods are covered with randomly oriented plate-like crystallites (Figure 2).

Figure 2: SEM images of Film A after annealing. Images were taken normal to the cleaved edge.

XRD analyses of Film A before and after thermal annealing are presented in Figure 3. The results indicate that the as-deposited films are polycrystalline Zn. After annealing, the films were oxidized to form polycrystalline ZnO films.

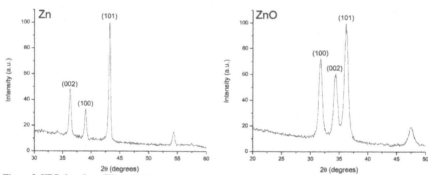

Figure 3: XRD data from Film A before (left) and after (right) annealing. The peaks identified are attributed to the crystal planes indicated for Zn (left) and ZnO (right).

Transmission electron microscopy performed on Film B after annealing revealed that the nanostructures within the film consist of polycrystalline trunks with single crystal nanorods growing off their surface (Figure 4). High-magnification bright field images reveal stacking of crystal planes within the single crystal nanorods. Electron diffraction images of the polycrystalline trunks and of the single crystalline nanorod are shown in Figure 4. The data confirms that the trunks are textured and polycrystalline, while the nanorods are single-crystalline. Scatter from the amorphous lacey carbon in the TEM grid likely produced the diffuse rings seen in the diffraction pattern for the single crystal nanorod (Figure 4).

64

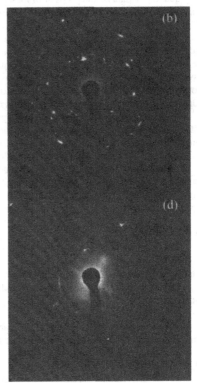

Figure 4: Bright field (left) and electron diffraction (right) images of polycrystalline ZnO trunk nanostructures (top) and single crystal ZnO nanorod (bottom). Typical nanorod circled in (a). Images taken from TEM grid prepared from Film B after annealing.

DISCUSSION

Porous Zn films were fabricated through an aperture and at a short throw-distance (40 mm to 50 mm) in an attempt to limit flux scatter and angular divergence on the substrate and thereby create the necessary self-shadowing conditions for columnar film growth commonly observed during GLAD. Instead of the expected columnar growth the films were porous consisting of randomly oriented crystallites. The packing density, aspect ratio and characteristic dimensions depend on the growth conditions. At large deposition angles highly-porous, large aspect ratio nanorods were grown (Figure 1 – Films A-D). Increasing chamber pressure during deposition appears to increase the film density (Film D). Deposition at normal incidence without an aperture produced denser films consisting of low-aspect ratio crystallites (Figure 1 – Film E, F). We expect films deposited at angles between 0° and 84° to show a gradual change in morphology from the low-aspect ratio crystallites at 0° to the high-aspect ratio nanorods at 84°.

Only qualitative data on the porosity of these films has been presented here. Surface area characterization of these films using gas absorption porosometry is ongoing. Films grown at longer throw distances, without an aperture (not presented) show a similar change in porosity with deposition angle, which suggests that a highly collimated flux is not necessary for the observed growth mode. The observed faceting of the crystallites suggests that the adatom surface mobility of Zn is sufficient to allow them to settle on the most energetically favorable crystal planes. Substrate heating may be important. It is unclear if the change in morphology at higher incident angles is due to geometric shadowing, a reduction in heating, or a combination of both. Further investigation is required to understand the mode of growth.

XRD analyses confirmed that post-deposition annealing was able to oxidize the Zn into ZnO without changing the film structure significantly. However, plate-like crystallites appear on annealed films. The smaller crystallites seen in Film B may be caused by Zn surface oxidization while the film was stored under ambient conditions.

We expected that any single crystal nanorods in the film would have a wurtzite crystal structure and that the nanorods would grow parallel to a crystal axis. Instead, TEM images (Figure 4) indicate that nanorods can grow in an off-axis direction. It is unclear if single-crystal growth occurs during deposition or annealing.

CONCLUSION

In conclusion, we have applied GLAD to the fabrication of porous crystalline Zn films. Changes in film morphology with deposition angle and pressure were presented and it appears that deposition angle affects the film porosity. The diversity of growth modes in sputtered Zn films presented in this study and others [4] indicate that further examination is required to understand sputtered Zn growth mechanisms.

Post-deposition annealing was used to form porous ZnO films. This method may allow for control over the morphology and surface area of porous nanostructured ZnO films. These films may useful for sensing and photovoltaic applications.

ACKNOWLEDGMENTS

Thanks to Dr. M.T. Taschuk for his constant encouragement, gentle guidance, and insightful advice, Dr. M.D. Fleischauer for his assistance with XRD, M.A. Summers and G. Braybrook for SEM operation, and H. Qian for TEM operation. This work was performed using funding from NSERC, iCore/Alberta Ingenuity and Micralyne.

REFERENCES

1. U. Ozgur, Y.I. Alivov, C. Liu, A. Teke, M.A. Reshchikov, S. Dogan, V. Avrutin, S.J. Cho, and H. Morkoc, J. Appl. Phys. 98, 041301 (2005).
2. Z.L. Wang, J. Phys: Cond. Matt. 16, R829-R858 (2004).
3. W. Chunrui and W.I. Park, Semi. Sci. & Tech. 20, S22-S34 (2005.
4. W. Chiou, W.Y. Wu, and J.M. Ting, Diamond and Rel. Matt. 12, 1841-1844 (2003).
5. K. Robbie, M. J. Brett, and A. Lakhtakia, Nature 384, 616 (1996).
6. M.M. Hawkeye, and M.J. Brett, J. Vac. Sci. & Tech. A. 25, 1317-1335 (2007).

Mater. Res. Soc. Symp. Proc. Vol. 1142 © 2009 Materials Research Society   1142-JJ05-39

Flame Synthesis of ZnO Nanostructures: Morphology and Local Growth Conditions

Fusheng Xu[1], Cassandra D'Esposito[1], Xiaofei Liu[1], Bernard Kear[2], and Stephen D. Tse[1]
[1]Department of Mechanical and Aerospace Engineering
[2]Department of Materials Science and Engineering
Rutgers University
Piscataway, NJ 08854, U.S.A.

ABSTRACT

Various ZnO nanostructures are produced from counterflow diffusion flames (CFDs), which include nano- wires, ribbons, rods, and networked structures. The synthesis is carried out at one atmosphere pressure in an open environment using a Zn-plated metal substrate. As described in a previous work, various ZnO nanostructures were obtained in inverse jet diffusion flames (IJDFs), given the appropriate local growth conditions (e.g. temperature and growth-related chemical species)[1]. However, the large radial gradients in IJDFs can result in non-uniformity of ZnO nanostructures, making it somewhat difficult to properly assess the growth conditions for specific ZnO nanostructures. To better correlate ZnO morphologies with local conditions, CFDs with quasi-one-dimensionality are employed, where the gradients vary mainly in the axial direction. The axial separation of fuel side and air side with respect to the reaction zone in CFDs also makes it easier to evaluate the roles of H_2O versus CO_2 versus O_2 in the synthesis of ZnO nanostructures. Laser-based diagnostics are used to map local chemical species concentrations and gas-phase temperature with ZnO growth morphology, helping to divulge the growth mechanisms. Results for ZnO nanostructures and their corresponding local conditions are then compared between methane and hydrogen flames to assess the roles of H_2O versus CO_2 versus O_2 on ZnO morphology.

INTRODUCTION

Nanostructured ZnO materials have attracted much interest in recent years due to their unique semiconducting, piezoelectric, and pyroelectric properties[2], as well as bio-safety and biocompatibility characteristics[3]. Possessing a wurtzite lattice structure of non-central symmetry[4] and a combination of three sets of fast growth directions (along with polar surfaces), ZnO can manifest itself in a wide range of diverse structures, such as nanowires, nanorods, nanoneedles, nanotubes, nanobelts, nanodisks, nanorings, and hierarchical and networked nanostructures[5]. Within this group, several techniques for the fabrication of one-dimensional ZnO nanowires have been explored using different starting materials, e.g. zinc metal powder, ZnO powder, and metal organic precursor.

Many approaches have been used to prepare nano-wires/rods, such as vapor-liquid-solid growth, solution-liquid-solid methods, template mediate growth, electron beam lithography, and scanning tunneling microscopy techniques. However, these methods can be technically complex, incompatible with complementary ZnO semiconductor technology, or incapable of pattern growth. As such, efforts to develop a simple method for the production of ZnO nanowires for multi-purpose applications without the above mentioned disadvantages are needed.

We utilize a combustion-based technique to directly synthesize monocrystalline zinc oxide nano-wires/rods from micro-sized metal grains at high rates under atmospheric conditions. This method is advantageous because the heat and chemical reactants required are inherently produced by the flame, and there is a demonstrated history of cost-effective scalability in such gas-phase systems (e.g. commercial production of titania, silica, carbon black, etc.). CDFs, which possess a quasi one-dimensional flame structure, are employed to correlate ZnO morphologies with local growth conditions, whereby local temperatures and chemical species concentrations can be tuned accordingly.

EXPERIMENT

The experimental setup consists of a CDF with a zinc-plated substrate probe inserted radially into the gas phase flame structure at six specified temperature conditions. Diluted-methane and hydrogen flames are compared, with their temperature profiles tailored to match. H_2O and CO_2, which are both present in the methane flame, can play different roles in the growth mechanism of ZnO nanostructures[1]. To distinguish the roles, hydrogen flames are examined because CO_2 is absent and only H_2O is present. The gas-phase temperatures and species concentrations are simulated computationally, using detailed chemical kinetics and transport properties, which are then compared to the measurements using spontaneous Raman spectroscopy (SRS), as seen in Fig. 1. An S-type thermocouple is employed to determine substrate temperatures, which can be different than gas-phase temperatures, due to radiative effects and conductive losses along the probe.

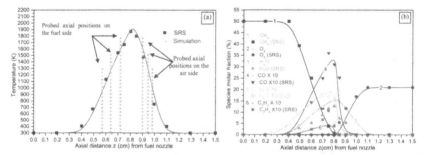

Figure 1. Methane flame structure as measured by SRS and compared with simulations: (a) temperature profile along the axial z direction and the probed positions; and (b) the mole fractions of major species along the axial z direction. Note: the hydrogen flame has the same temperature profile as shown in (a).

The morphologies of as-grown nanostructures are examined using field emission scanning electron microscopy (FESEM, LEO Zeiss Gemini 982). Elemental analysis is conducted using energy dispersive x-ray spectroscopy (EDX) attached to the FESEM. Structural features of the nanomaterials are investigated using high resolution transmission electron microscopy (HRTEM, TOPCON 002B), along with selected area electron diffraction (SAED).

DISCUSSION

Figures 2 and 3 show the various structures obtained at z=9mm (T=1600K), on the air side, in the methane and hydrogen diffusion flames, respectively. While the temperatures are the same, the species concentrations differ. There are 3.92% CO_2, 9.43% H_2O, and 7.59% O_2 in the methane flame, versus 11.5% H_2O and 8.82% O_2 in the hydrogen flame at the given location. Nanorods with a diameter of approximately 450 nm are shown in Fig. 2(a), displaying uniform hexagonal cross-sections along the growth direction. "Welded" joints of nanorods are observed in Fig. 2(b), which may be formed by a sintering process at the high local temperature[6]. Nanorods with sharp tips are visible in Fig. 2(c), which may result from a local decrease in ZnO vapor supply as source material is consumed[7]. The "nanonails"[3,8] in Fig. 2(d) exhibit a gradually-decreasing cross section from the top to bottom of the structure. The growth of "nanonails" may be due to better absorption of incoming ZnO vapor at the top than at the bottom due to high packing density[8]. Nanorods with sharp tips arranged into flower-like patterns[9] are shown in Fig. 2(e). The nanorods grow out of the substrate radially and aggregate together to form a flower-like pattern, which are likely comprised of hexagonal facets[8]. Figure 2(f) shows complicated structures that are developed by multiple growths of nanorods. Tetrapod-shaped structures, as shown in Fig. 2(g), are comprised of four rod-shaped arms situated at tetrahedral angles from a central core[10]. These four arms have a hexagonal cross section with uniform diameter and length. It is generally proposed that the presence of a tetrahedral zinc blended core nucleate the growth of four cylindrical arms[11]. Spatially on the substrate, the zones of nanorod and "nanonail" growth are each larger than 100 μm with a cross-over section of about 20 μm consisting of a mixture of the two nanostructures.

Nanorods composed of smaller, more complex nanostructures can be seen in Fig. 3(a). An area of nanoribbons that are less than 500 nm wide are shown in Fig. 3(b). In Fig. 3(c), longer "nanonails" are produced from the hydrogen flame than from the methane flame. The nanostructures shown in Fig. 3(d) are similar to those seen in Fig. 2(f).

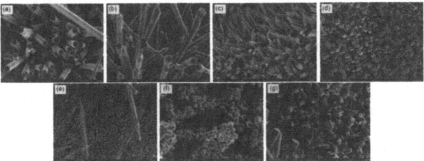

Figure 2. FESEM images of nanostructuress produced in the CH_4 flame on the oxidizer side where T=~1600K.

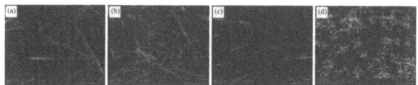

Figure 3. FESEM images of nanostructures produced in the H_2 flame on the oxidizer side where T=~1600K.

Figures 4 and 5 display FESEM images of ZnO nanostructures obtained at the axial position z=9.4 (T=1300K), on the oxidizer side of the methane and hydrogen flames, respectively. The local species conditions for the methane flame are 2.49% CO_2, 7.17% H_2O, and 11.9% O_2, while for the hydrogen flame are 8.64% H_2O and 12.6% O_2. A high-magnification image of nanorods in Fig. 4(a) clearly shows hexagonal facets. Radially away from the flame centerline, a small quantity of ZnO nanoribbons are found to be intermixed with nanowires of ~20 nm diameter, as displayed in Fig. 4(b),. These nanoribbons are ~50 nm wide and ~10 nm thick. Tower-like structures, consisting of many individual hexagonal plates stacked up layer upon layer[12], are observed in Fig. 4(c). Similarly, chain-like structures, composed of many ZnO sharp cones connected to one another, are shown in Fig. 4(d). Each cone is characterized by an obvious hexagonal bottom and a sharp tip.

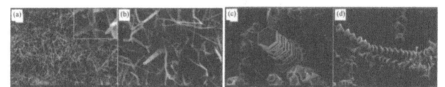

Figure 4. FESEM images of nanomaterials from the CH_4 flame on the oxidizer side where T=~1300K:
(a) nanorods, (b) nanoribbons, (c) tower-like structure, (d) chain-like structure.

Figure 5(a) shows images of large nanorods a few hundred nanometers in diameter. Nanostructures similar to the tetrapod-shaped structures of Fig. 2(g), which have arms extending from the center of the structure, are evident in Fig. 5(a).

Figure 5. FESEM images of nanomaterials from the H_2 flame on the oxidizer side where T=~1300K.

In order to study the effects of oxygen on ZnO nanostructure growth, an axial position of z=9.8 mm with a temperature of 1000K is probed. These conditions are characterized by 1.47% CO_2, 5.11% H_2O, and 1.51% O_2 for the methane flame, and 6.23% H_2O and 1.54% O_2 for the hydrogen flame. At this position, the main structures produced are micro-sized columns as seen in Fig. 6(a). They possess irregular hexagonal cross sections and are at least 10 μm in diameter and are most likely nucleated from large Zn droplets. A small amount of nanosheets with a thickness of ~50 nm and of various shapes are observed radially away from the flame centerline, which can be observed in Fig. 6(b). Figures 6(c) and 6(d) are comprised of large nanorods ~300 nm in diameter.

Figure 6. FESEM images of nanomaterials from the axial position on the oxidizer side where T=~1000K: (a) microsized columns/chunks (CH_4), (b) nanosheets (CH_4), (c) and (d) nanorods from the H_2 flame.

The fuel side of the CDF, where O_2 does not exist, is investigated to study the role of H_2O on ZnO growth; see Fig. 1. Matching the temperatures probed on the air side of the CDF, three axial positions are examined; see Fig. 1. Figure 7 displays the nanostructures obtained from these experiments. As expected, perfect hexagonal nanorods are observed at T=~1600 K with a uniform diameter of ~450 nm, as shown in Fig. 7(a). By comparing the growth conditions on the oxidizer side where the gas phase temperature is the same, we find that temperature plays a key role in determining ZnO nanostructures, where nanorods are almost identical (in terms of diameter) despite the differences in O_2 concentration. Nanorods produced at T=~1300 K have an average diameter of ~150 nm and lengths up to 30 μm (Fig. 7(b)). At z=5.7 mm (T=1000K), Fig. 7(c) shows thin nanowires ranging from 30-100 nm in diameter. The local species conditions are 2.44% CO_2 and 6.93% H_2O. Away from the flame centerline at this axial position, a very small quantity of nanoribbons can be observed in Fig 7(d). These nanoribbons are ~150 nm wide and ~25 nm thick and can be up to several tens of micrometers in length.

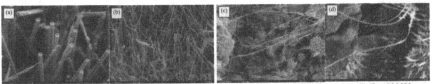

Figure 7. ZnO nanomaterials from the fuel side. (a) T=~1600K, (b) T=~1300K, (c) and (d) T=~1000K.

In theory, for the quasi-one-dimensional CDF, the same morphology of ZnO nanomaterials should be produced across the radial direction at an axial position on both the oxidizer and fuel sides. However, our results show that to some extent, different morphologies and structures appear along the radial directions for given axial positions. For example, at T=1600K (z=9.0mm) on the oxidizer side, although major products are nanorods with hexagonal cross sections (Fig. 2), there is some quantity of tetrapod structures and nanorods with a sharp tip. A slight radial temperature gradient, resulting from conductive losses along the length of the substrate probe, likely plays a role. The slight growth-condition changes induced by this factor may be enough to change the structures and morphologies of the ZnO nanomaterials, since they are very sensitive to the local growth environment. Wang et al.[13] reported in their work that the temperature, vapor flow, and the availability of the Zn-O vapor could affect the morphologies of as-prepared ZnO structures. The work by Hu et al.[12] concluded that ZnO morphologies are closely related with the reaction temperature, oxygen partial pressure, and flow rate. Gao et al.[14] also found that the local temperature and surface diffusion rate have an influence on ZnO nanostructures. Nevertheless, in general, the ZnO nanomaterials produced in CDFs are quite uniform (even in the radial direction)—namely, one structure/morphology constitutes the major product (~70% by volume) of harvested material at a given axial position. At the same time, the slight gradients in the radial direction seem to have no influence on carbon nanotube growth in CDFs, as reported in Xu et al.[15], with reasonably uniform morphologies CNTs harvested along the probe in terms of diameter, yield, and alignment. The sensitivity of ZnO nanostructures to a slight radial gradient is likely due to a liquid Zn layer on the surface of the probe which is susceptible to Marangoni effects. A liquid Zn layer can exist on the substrate surface due to the low melting point of Zn (692K). The surface tension of this liquid layer is temperature dependent and would decrease with increasing temperature[16]. Such a gradient in surface tension causes liquid to flow away from regions of low surface tension to those of high surface tension, which pull more strongly on the surrounding liquid. This mechanism can intermix adjacent regions, likely affecting the local nucleation of ZnO nanostructures, the preferential faceting of nanostructures, the local reaction of Zn with H_2O/CO_2, and the absorption of ZnO vapor, resulting in the different nanostructures and morphologies of final as-grown ZnO, along the radial length of the probe.

CONCLUSIONS

We demonstrate the synthesis of single-crystalline ZnO nanowires grown directly from metal substrates at high rates in a flame process. Comparisons of growth morphology between methane and hydrogen flames help to assess growth mechanisms involving H_2O versus CO_2. The growth mechanism appears to be vapor-solid based, with key parameters being radical species present, oxidizer and water-vapor concentrations, substrate temperature, and gas-phase temperature.

ACKNOWLEDGMENTS

This work is sponsored by the Army Research Office (grant W911NF-08-1-0417), the National Science Foundation (grant NSF-CBET-0755615), and the Office of Naval Research (grant N00014-08-1-1029). The authors appreciate the help of Mr. John Petrowski with the experiments.

REFERENCES

1. F. Xu, X. Liu, S.D. Tse, F. Cosandey, B.H. Kear, Flame synthesis of zinc oxide nanowires, Chem. Phys. Lett. 449:175-181 (2007).
2. Z.L. Wang, Nanostructure of zinc oxide, Materials Today 7: 26 (2004).
3. G.C. Yi, C. Wang, W.I. Park, ZnO nanorods: synthesis, characterization and applications, Semicond. Sci. Technol. 20:S22 (2005).
4. Z.L. Wang, Zinc oxide nanostructures: growth, properties and applications, J. Phys.: Condens. Matter 16:R829 (2004).
5. C.X. Xu, X.W. Sun, Z.L. Dong, G.P. Zhu, Y.P. Cui, Zinc oxide hexagram whiskers, Appl. Phys. Lett. 88:093101 (2006).
6. P. X. Gao, C. S. Lao, W. L. Hughes, Z.L. Wang, Three-dimensional interconnected nanowire networks of ZnO, Chem. Phys. Lett. 408:174-178 (2005).
7. P.X. Gao, Z.L. Wang, Nanopropeller arrays of zinc oxide, Appl. Phys. Lett. 84(15):2883-2885 (2004).
8. J. Y. Lao, J. Y. Huang, D. Z. Wang, and Z. F. Ren, ZnO Nanobridges and Nanonails, Nano Lett. 3(2): 235-238 (2003).
9. W. Bai, K. Yu, Q. Zhang, F. Xu, D. Peng, and Z. Zhu, Large-scale synthesis of ZnO flower-like and brush pen-like nanostructures by a hydrothermal decomposition route, Materials Letters 61:3469-3472 (2007).

10. Y. Dai, Y. Zhang, and Z. L. Wang, The octa-twin tetraleg ZnO nanostructures, Solid State Communications 126(11): 629-633 (2003).
11. M.C. Newton, P.A. Warburton, ZnO tetrapod nanocrystals, Materials Today 10(5):50-54 (2007).
12. P.A. Hu, Y. Q. Liu, L. Fu, X.B. Wang, D.B. Zhu, Controllable morphologies of ZnO nanocrystals: nanowire attracted nanosheets, nanocartridges and hexagonal nanotowers, Appl. Phys. A, 80:35-38 (2005).
13. F. Wang, L. Cao, A. Pan, R. Liu, X. Wang, X. Zhu, S. Wang, B. Zou, Synthesis of tower-like ZnO structures and visible photoluminescence origins of varied-shaped ZnO nanostructures, J. Phys. Chem, C, 111:7655-7660 (2007).
14. P.X Gao, Z.L Wang, Appl. Phys. Lett. 84:2883 (2007).
15. F. Xu, H. Zhao, S.D. Tse, Carbon nanotube synthesis on catalytic metal alloys in methane/air counterflow diffusion flames, Proc. Combust. Inst. 31:1839–1847 (2007).
16. W.J. Moore, Physical Chemistry, 3rd ed. Prentice Hall (1962).

Mater. Res. Soc. Symp. Proc. Vol. 1142 © 2009 Materials Research Society   1142-JJ05-59

Chemistry of Reverse Micelles: A Versatile Route to the Synthesis of Nanorods and Nanoparticles

Tokeer Ahmad[1]*, Ashok K. Ganguli[2], Aparna Ganguly[1,2], Jahangeer Ahmed[2], Irshad A. Wani[1], Sarwari Khatoon[1]

[1]Department of Chemistry, Faculty of Natural Sciences, Jamia Millia Islamia, New Delhi – 110025, INDIA
[2]Department of Chemistry, Indian Institute of Technology, Hauz Khas, New Delhi 110016, INDIA

ABSTRACT

Nanostructured wires and rods are expected to have interesting optical, electrical, magnetic and mechanical properties as compared to micron sized whiskers and fibers. We have explored a versatile route for the synthesis of nanorods of transition metal (Cu, Ni, Mn, Zn, Co and Fe) oxalates, succinates of few metals (Co and Fe) and SnO_2 nanoparticles using reverse micelles. The aspect ratio of the nanorods copper oxalate could be modified by changing the solvent. The aspect ratio of the cobalt oxalate nanorods could be modified by controlling the temperature. The nanorods of metal (Cu, Ni, Mn, Zn, Co and Fe) oxalates were found to be suitable precursors to obtain a variety of transition metal oxide nanoparticles. Our studies show that the grain size of CuO nanoparticles is highly dependent on the nature of non-polar solvent used to initially synthesize the oxalate rods. All the commonly known manganese oxides could be obtained as single phases from the manganese oxalate precursor by decomposing in different atmosphere (air, vaccum or nitrogen). In order to see the templating effect of the ligand we have changed the dicarboxylate ligand by using succinate (4 carbon chain) instead of oxalate (2 carbon chain). We found spherical nanoparticles for iron succinate where the oxidation state of Fe is in +3. Shorter rods of cobalt succinate were observed. Monophasic tin dioxide (SnO_2) nanoparticles with an average size of 6 - 8 nm was obtained at 500°C by the reverse micellar route using liquor NH_3 as precipitating agent.

INTRODUCTION

Nanoscience is among the most challenging area of current scientific and technological research because of the variety of interesting changes in the optical, magnetic, electrical and catalytic properties at nano-dimension. Superior properties have been demonstrated for a broad range of such materials when the size has been reduced to nano-dimension. These nanoscale materials can be defined as those whose characteristic length (at least one length) lies between one and hundred nanometers. The properties of nanomaterials are significantly different from individual atoms or molecules and from bulk materials. Microemulsion or reverse micellar approach is one of the versatile methods for the synthesis of these nanomaterials [1-6]. In 1982, Boutonnet et al reported the first synthesis of material using reverse micelles [7]. Metal, metal oxide, ceramics, semiconductor, polymers have been synthesized using this method [8,9]. The structure of reverse micelle can be elucidated as a nanosized droplet of polar liquid phase capped with the surfactant molecules, acting as the nanoreactor, dispersed in organic phase. Microemulsion is thus defined as an optically transparent and thermodynamically stable system of oil, water and amphiphiles. It is the interior of the reverse micelle that acts as a template for the growing nuclei into narrowly dispersed or perfectly uniform nanoparticles. The most important aspect of using the reverse micelles is to control the size of the nanoparticle. Apart from stirring time and temperature, type of surfactant, solvent, co-surfactant, concentration of reagents and the molar ratio of water to surfactant used are most crucial. The synthetic methodology involves the preparation of two individual microemulsions, incorporating the different ionic contents. On mixing, the nucleation occurs on the micelle edges and the growth then occurs around this nucleation point. It is thus important to note at this point that these microemulsion systems are dynamic i.e. micelles coalesce and decoalesce for homogenous mixing. The rate limiting step for particle growth is the intermicellar exchange of the aqueous core [10]. Based on this we chose a cationic surfactant, CTAB and synthesized a variety of transition metal dicarboxylates. This review describes the formation of nanorods of transition metal oxalates and succinates in the presence of the cationic surfactant. The nanometer-diameter wires, which can be grown from a variety of materials and reach tens of micrometers in length, have one major advantage - unlike nanotubes, their chemistry is relatively easy to tailor. Numerous processes are being developed aiming at the synthesis of nanorods and nanowires. Among them, the vapour-liquid-solid (VLS) growth [11, 12], solution-liquid-solid (SLS) method [13], template-mediated growth [14, 15], electron-beam lithography [16] and other methods of nanowire production are known [15-17]. Single crystal oxide nanowires and nanobelts could be grown directly through evaporation-condensation process under vacuum at elevated temperatures [18]. VLS growth of oxide nanorods and nanowires is restricted to systems that can form a eutectic liquid with the catalyst at the growth temperature [19]. Various chemical, electrochemical and physical deposition techniques have been employed to create oriented arrays of ZnO nanorods and nanowires that include physical vapor deposition [20], vapor phase transport [21], chemical vapor deposition [22, 23], surfactant assisted hydrothermal method [24, 25], and soft solution method [26]. Ours is a relatively easy method of synthesizing one dimensional nanostructures with surfactant mediated synthesis. The thermal decomposition of these transition metal oxalate rods at low temperatures of 450-500°C yield homogenous oxide nanoparticles. Our methodology using reverse micelles results in the formation of highly uniform rods of the metal oxalates and not nanoparticles as obtained in direct chemical reaction of metal salt and oxalic acid [27-29].

SnO_2 is a wide-band gap (3.6 eV) n-type semiconductor, which is widely used for gas sensors. It has been utilized in various applications such as gas sensors [30], microelectronics [31], solar cells [32] and photoelectrochemistry [33]. The compound has also been examined as possible electrode material for lithium cells [34] and photocatalysts [35]. As an n-type semiconductor, tin dioxide shows very high sensitivity towards reducing gases such as H_2, CO, hydrocarbon, and alcohol.

EXPERIMENTAL

The qualitative explanation for the formation of nanorods is the assembly of surfactant molecules on the linear structure of the metal oxalates [36, 37] which probably leads to the formation of nanorods. The syntheses of nanorods has been carried out by mixing the two optically transparent micro-emulsions (I and II).Micro-emulsion I is composed of cetyltrimethylammonium bromide (CTAB) as the surfactant, n-butanol as the co-surfactant, isooctane (or n-octane in some cases) as the hydrocarbon phase and aqueous phase containing

0.1M solution of metal ion (copper nitrate, nickel nitrate, zinc nitrate, manganese acetate, ferrous nitrate and cobaltous nitrate) solution as the aqueous phase. Microemulsion II comprised of the same constituents as above except for having 0.1M ammonium oxalate instead of metal ions as the aqueous phase. These two micro-emulsions were mixed slowly and stirred overnight. The product was separated from the apolar solvent and surfactant by centrifuging and washing it with 1:1 mixture of methanol and chloroform. The precipitate was dried at room temperature. Similar to the above synthesis, metal succinate complexes have also been synthesized in order to study the effect of the long chain dicarboxylate ligand on the morphology of the product formed.

Metal oxide nanoparticles were obtained by the decomposition of the oxalate precursor at 450°C to 500°C for 6h, on the basis of the thermogravimetric studies. CuO nanoparticles were obtained by the thermal decomposition of copper oxalate monohydrate, synthesized by using both isooctane and n-octane as the non-polar medium. The manganese oxalate nanorods were subjected to a careful thermal decomposition to yield nano-particulate manganese oxides. Anhydrous manganese oxalate was obtained by heating the hydrated precursor at 150°C .The anhydrous precursor was then heated in air at 450°C for 6h to obtain of α-Mn$_2$O$_3$ and under nitrogen at 500°C (10h), Mn$_3$O$_4$ was obtained. MnO was obtained by decomposing manganese oxalate in sealed quartz tube at a pressure ~10^{-5} torr. The oxides of Ni and Zn were obtained by the decomposition of metal oxalates in air at 450°C for 6h. Nanorods of cobalt and iron oxalate dihydrate were decomposed under different atmospheric conditions (air, helium and hydrogen) to obtain the nanoparticles of various cobalt oxides (Co$_3$O$_4$, CoO and metallic Co nanoparticles) and iron oxides (Fe$_3$O$_4$ and Fe$_2$O$_3$).

SnO$_2$ nanoparticles have been synthesized using the microemulsion method. The precipitate was dried in an oven at 60°C for 2 hrs. The powder x-ray diffraction pattern shows an amorphous nature at 60°C. The precursor on heating in air at 500°C for 5 hrs led to the formation of monophasic SnO$_2$.

Powder X-ray diffraction studies (PXRD) were carried out on a Bruker D8 Advance X-ray diffractometer using Ni filtered Cu Kα radiation. The grain size was calculated using the Scherrer's formula [38] (t= 0.9λ/Bcosθ) where t is the diameter of the grain, λ is the wavelength of the radiation (λ for Cu Kα is 1.5418Å) and B is the line broadening which is measured from the broaden peak at full width at half maxima (FWHM) and calculated by the Warren's formula [39, 40]; $B^2 = (B_M^2 - B_S^2)$ where B$_M$ is the full width at half maxima of the sample and B$_S$ is the fullwidth at half maximum of the standard quartz with a grain size of around 2μm. The cell parameters were determined using a least square fitting procedure on all reflections using quartz as the external standard.

Thermogravimetric (TGA) and differential thermal analysis (DTA) was carried out using Perkin-Elmer TGA and DTA system on well ground samples in flowing nitrogen atmosphere with a heating rate of 5°C/min. FTIR spectra of the nanorods of copper and nickel oxalate powders (using KBr disks) were recorded in the range from 4000 - 400 cm^{-1} using a Nicolet Protege 460 (FTIR) spectrometer. Transmission electron microscopic (TEM) studies were carried out using JEOL JEM 200CX model and Technai G^2 20 (FEI) model both operated at 200kV.TEM specimens were prepared by dispersing the powder in acetone by ultrasonic treatment, dropping onto a porous carbon film supported on a copper grid, and then drying in air.

RESULTS AND DISCUSSION

[A] Structural analysis of transition metal (Cu, Ni, Mn, Co, Zn and Fe) oxalate and (Co – Fe) succinate nanorods

Using the microemulsions containing the metal ions and oxalate ions, we could synthesize metal oxalate at room temperature. The powder X-ray diffraction confirmed the formation of pure phase metal oxalates. The PXRD pattern for the orthorhombic Cu and Co oxalate system has been shown in Fig 1a and 1b respectively while monoclinic Ni, Mn, Fe and Zn oxalates has been shown in Fig1 (1c, 1d, 1e and 1f). Contrary to the oxalates, the metal succinates obtained were amorphous and showed no improvement in the crystallanity on heating at higher temperatures.

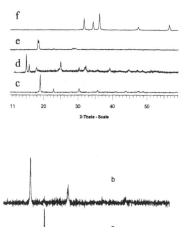

Figure 1. PXRD patterns of (a) Cu (b) Co (c) Ni (d) Mn (e) Fe and (f) Zn oxalate nanorods.

[B] Thermal studies of transition metal oxalate and dicarboxylate nanorods

The presence of water molecule in all oxalate nanorods has been confirmed using thermo gravimetry and differential thermal analysis (TGA/DTA). All the experiments were carried out in flowing nitrogen. The number of water of crystallization associated with each of these oxalates as calculated from the TGA data and the temperature at which they decompose to give the anhydrous product and then to its corresponding oxide has been inferred from the TGA data.

[C] Structural analysis of transition metal oxide nanoparticles obtained by the thermal decomposition of transition metal oxalate nanorods

Based on the Thermo gravimetric data, the oxalate precursors were decomposed at different temperatures under various conditions to obtain the corresponding oxides. The indexed PXRD pattern for the oxides showing the prominent peaks has been shown in figure 2. Monophasic zinc oxide nanoparticles were obtained by the decomposition of nanorods of zinc oxalate and the PXRD pattern (Fig 2a) could be indexed on the basis of a hexagonal cell with the refined lattice parameters of 'a' = 3.2498(4) Å and 'c' = 5.209(1) Å which matches with the reported pattern of ZnO [41]. However there is a small increase in the 'c' parameter of our nanocrystalline ZnO compared to the bulk value of 5.203(1) Å [42]. This may be explained due to size effect on the lattice parameters. The decrease in the grain size can cause lattice expansion [43, 44]. This has been explained by the increase in number of surface atoms of the nanoparticles leading to atomic disorder and reduced coordination of the surface atoms causing the lattice expansion. Nanoparticles of pure NiO and CuO (using iso-octane as the non-polar solvent) were obtained by heating the corresponding oxalates at 450°C as shown in the PXRD (Fig 2b and 2c respectively). The refined unit cell parameters of copper oxide are ['a' = 4.684(2) Å, 'b' = 3.4256(7) Å, 'c' =5.126(3) Å, β = 99.50°].

Nanorods of manganese oxalate were decomposed under different conditions to obtain oxides in various oxidation states. Pure phase of Mn_2O_3 (Fig 2d, cubic, 'a' = 9.425(3) Å), Mn_3O_4 (Fig 2e, tetragonal, 'a' = 5.7631(9) Å, 'c' = 9.458(3) Å) and MnO (Fig 2f, cubic, 'a' = 4.4407(5) Å) nanoparticles were obtained by heating in air, nitrogen and vacuum respectively. Nanorods of cobalt oxalate dihydrate were also decomposed under different conditions (various temperatures and atmospheres) to obtain nanoparticles of cobalt oxides in different oxidation states. On heating the oxalate precursor at 500°C for 12h in air, Co_3O_4 nanoparticles (Fig 1g)with cubic structure and refined lattice parameter of 'a' = 8.089(5) were obtained. However, a mixture of cobalt oxide (CoO) and Co nanoparticles was obtained from the oxalate precursor when heated in nitrogen or helium atmosphere at 500°C for 10h.

Fe_2O_3 nanoparticles have been obtained by decomposing the oxalate precursor at 500°C in air (Fig 2h). It was indexed on the basis of a rhombohedral structure (JCPDS No. # 01-1053) with the refined hexagonal lattice parameters of 'a' = 5.021(4) Å and 'c' = 13.682(7) Å. Another oxide Fe_3O_4 was obtained by decomposing iron oxalate in sealed quartz tube at a pressure ~10^{-5} torr. Fig 2i shows the formation of monophasic Fe_3O_4, which crystallizes in the cubic structure with refined unit cell parameter of 'a' = 8.381(1) Å which is close to the reported value (JCPDS No. 01-1111).

[D] Particle size analysis of transition metal oxalate nanorods
The dimensions of transition metal oxalate nanorods as obtained by the transmission electron microscopic (TEM) studies. The average dimensions of the nanorods of copper oxalate monohydrate were 130 nm (diameter) and 480 nm (length) and are more compacted corrugated sheet-like structures (Fig.3a). These nanorods were obtained with iso-octane as the organic solvent. Interestingly, when the non-aqueous solvent was changed to n-octane, rods of slightly different dimension were observed (100 nm in diameter and 640 nm in length) (Figure 3b) and appear to be more fibrous. Thus it appears that the solvent plays an important role in controlling the aspect ratio of these nanorods [1].
The average dimensions of the highly uniform and monodisperse nickel oxalate dihydrate nanorods was found to be 250 nm (diameter) and 2.5 μm (length) (Fig 3c) [1]. TEM studies for anhydrous manganese oxalate show the formation of nanorods (Fig 3d) with average dimensions of 100 nm (diameter) and 2.5 μm (length) [45]. Zinc oxalate dihydrate forms nanorods (Fig 3e)

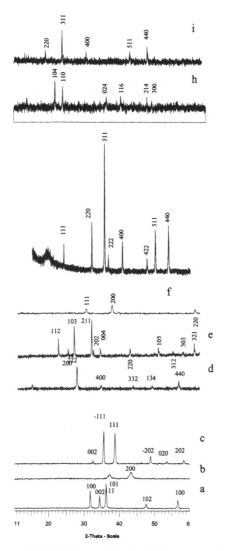

Figure 2. PXRD pattern of (a) ZnO (b) NiO (c) CuO (d) Mn2O3 (e) Mn3O4 (f) MnO (g) Co3O4 (h) Fe2O3 and (i) Fe3O4 nanoparticles

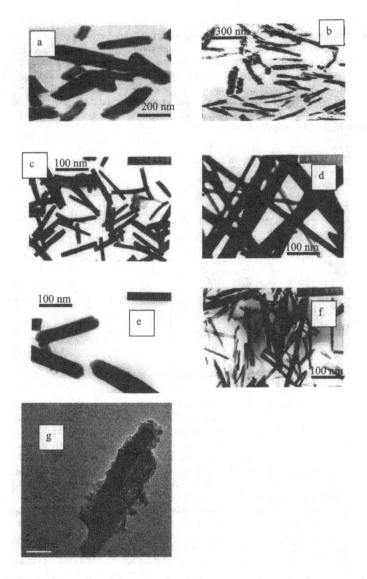

Figure 3. TEM micrographs of the nanorods of (a) copper oxalate using isooctane solvent (b) copper oxalate using n-octane solvent (c) nickel oxalate (d) Mn-oxalate (e) Zn-oxalate (f) cobalt oxalate (g) iron-oxalate.

with average dimensions of 120 nm (diameter) and 600 nm (length) [2]. TEM studies of cobalt oxalate dihydrate have been performed systematically after heating at various temperatures to see the effect of temperature on the morphology of nanorods of dihydrate and anhydrous cobalt oxalate obtained at room temperature and 185°C respectively (fig 3f). The average diameter of the nanorods remained nearly unchanged at 300 nm till 150°C while the diameter and length of the anhydrous rods was become to 100 nm and 1 μm at 185°C. These nanorods were quite long as compared to nanorods of copper, nickel, manganese and zinc oxalate. The average dimensions of the iron oxalate dihydrate nanorods were 70 nm in diameter and 470 nm in length as shown by the low resolution TEM image (Figure 3g).

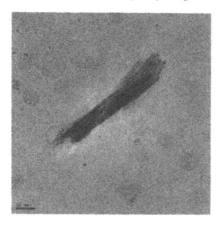

Figure 4. TEM micrograph of cobalt succinate sesqihydrate nanorods.

TEM studies of cobalt succinate dihydrate showed rods of ~ 500 nm length and 50 nm diameter (Fig 4) while for iron (III) succinate synthesized from Fe(NO₃)₃.9H₂O showed spherical particles of size ~ 150-200 nm (Fig. 5a). Mössbauer study was carried out to confirm the oxidation state (O.S.) of Fe. The isomer shift values are consistent with trivalent iron in octahedral geometry. It is thus expected that due to the +III O.S. of the metal ion, the ratio of metal ion to ligand required for charge neutralization would be 2:3. This ratio of metal to ligand results in the formation of spherical particles instead of a rod- like structure, which would normally require a 1:1 ratio of metal ion: ligand as shown in earlier studies on metal oxalates [1]. In an attempt to further investigate the role of oxidation state of the metal ion, we used $FeCl_2$ .4H₂O as the starting reagent instead of Fe(NO₃)₃. The TEM studies show smaller spherical particles of size ~ 20-30 nm (Fig5b). Mössbauer studies confirmed the presence of trivalent O.S. of Fe. The large quadrupole splitting (QS) values show significant distortion of the octahedra around Fe atom. From the above studies it appears that though the metal ion was initially in the +2 O.S. (FeCl₂), it gets oxidized during the preparation of the microemulsion or during the reaction when the two microemulsions were mixed.

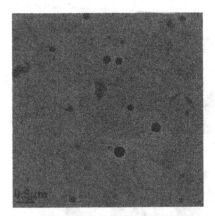

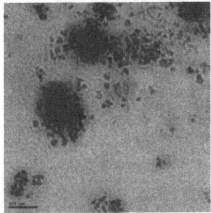

Figure 5. (a) TEM of iron succinate synthesized using FeCl₃ and (b) TEM of iron succinate synthesized using FeCl₂.4H₂O at room temperature.

[E] Particle size analysis of transition metal oxide nanoparticles

The oxalate nanorods on decomposition around 450 to 500°C led to the formation of oxide nanoparticles. Transmission electron micrographs of the CuO nanoparticles (synthesized using isooctane as the non-polar solvent) show particles with size in the range of 25-30 nm (Fig 6a). In contrast, much larger particles of the order of 80-90 nm were obtained from the rods synthesized using n-octane (Fig 6b). The particles were nearly spherical and showed minimal agglomeration. The grain size of nanostructured nickel oxide was evaluated from X-ray line broadening (using Scherrer's equation) and was found to be 30 nm which is in close agreement with the TEM studies with ~ size of 25 nm (Fig 6c).

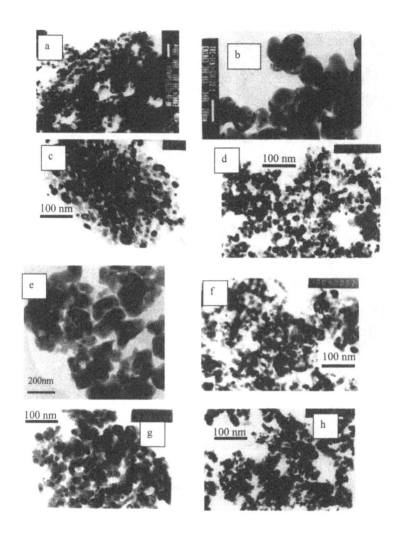

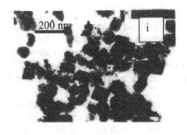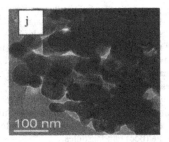

Figure 6. TEM micrograph of nanoparticles (a) CuO (using isooctane solvent) (b) CuO (using n-octane solvent) (c) NiO (d) MnO (e) Mn_3O_4 (f) Mn_2O_3 (g) ZnO (h) Co_3O_4 (i) Fe_3O_4 (j) Fe_2O_3.

X-ray line broadening studies of MnO showed a grain size of 32 nm close to what has been observed in the TEM studies (28 nm) (Fig 6d). However, there is a slight increase in particle size of Mn_2O_3 nanoparticles. From the X-ray diffraction studies, the grain size was evaluated to be 45 nm and that from TEM studies ~ 50 nm (Fig 6f), which corroborates the results, obtained from X-ray line broadening studies. Much larger grains (100 nm) were found in the case of Mn_3O_4 nanoparticles (Fig 6e). It may be noted that there is a gradual increase in particle size from MnO to Mn_3O_4. The grain size of ZnO nanoparticles as evaluated from X-ray line broadening studies was found to be 45 nm. TEM micrograph (Fig 6g) showed spherical particles with close similarity to the XRD results with ~ size 55 nm. The grain size of Co_3O_4 was found to be 28 nm using X-ray line-broadening studies, while TEM micrograph shows a slightly higher size of 35 nm (Fig.6h).

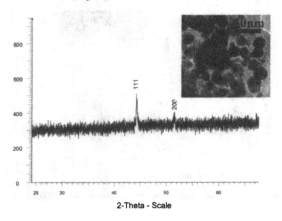

Figure 7. PXRD pattern of Co nanoparticles obtained by heating cobalt succinate at 650 °C under nitrogen. Inset shows TEM micrograph of Co nanoparticles.

85

For iron oxide nanoparticles (Fe_2O_3 and Fe_3O_4) with different oxidation states, the line broadening studies gave nearly the same result. The average grain size was found to be 47 nm for Fe_2O_3 and 44 nm for Fe_3O_4. Fe_3O_4 nanoparticles have faceted (cuboidal) grains as shown by TEM micrograph of sizes in the range of 60-70 nm (Fig 6i). On the contrary TEM micrograph in Fig 6j shows nearly spherical nanoparticles of Fe_2O_3 of average grain size ~50 nm. Some spherical particles join to form a Y-junction.

We have also carried out the decomposition of the succinates in the presence of nitrogen. At 650 °C, cobalt succinate dihydrate decomposed to yield pure Co nanoparticles (Fig.7) with size ranging between 10-40 nm (inset of Fig.7) while iron succinate was found to decompose at 800 °C to pure α- Fe (bcc) nanoparticles.

[F] Tin dioxide nanoparticles

Figure 8 shows the powder X-ray diffraction patterns of SnO_2 nanoparticles which are obtained by the reverse micellar route. All the reflections in the pattern could be indexed on the basis of a tetragonal cell (cassiterite) reported for SnO_2 (JCPDS # 21-1250). The reflections are markedly broadened, which indicate that crystallite size of SnO_2 nanoparticles is small. The average size was found to be 6-8 nm from TEM studies (inset of Fig. 8) which matches with the crystallite size calculated by the Scherrer's equation.

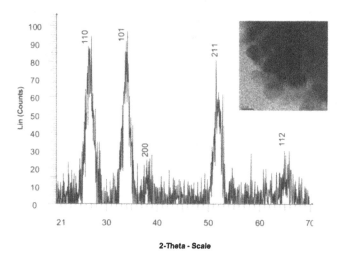

Figure 8. PXRD pattern of SnO_2 nanoparticles. Inset shows TEM micrograph of SnO_2 nanoparticles.

CONCLUSIONS

Synthesis of nanorods of copper oxalate monohydrate, nickel, manganese, zinc, cobalt and iron oxalate dehydrate and cobalt and iron succinate have been successfully accomplished through the reverse micellar route. TEM micrographs show the dependence of aspect ratio of the nanorods (copper oxalate monohydrate) on the nature of non-polar solvent. Larger (80-90 nm) copper oxide nanoparticles were observed when straight chain hydrocarbon (n-octane) was used as compared to 25-30nm sized particles obtained from iso-octane (branched chain hydrocarbon). Copper oxalate nanorods appear to be corrugated and crystallize in the monohydrate phase. Apart from it, all the other metal oxalates crystallize as dihydrates and are smooth. The aspect ratio of the nanorods of cobalt oxalate was found to be temperature dependent. The synthetic route developed for the synthesis of oxalate nanorods is versatile and a large number of monophasic metal oxalate nanorods can be obtained. Controlled thermal treatment of the transition metal oxalate /succinate under different atmospheric conditions yielded metal and metal-oxide nanoparticles. A change in the morphology of metal carboxylates was also observed when succinate was used instead of oxalate. Spherical nanoparticles were formed for iron succinate opposed to the nanorods we got for iron oxalate. We also find a significant change in particle size for the spherical nanoparticles of iron succinate as we change the reactant metal ion from $Fe(NO_3)_3.9H_2O$ (150-200 nm) to $FeCl_2.4H_2O$ (20-30 nm).

ACKNOWLEDGMENTS

TA and AKG thank the Department of Science & Technology, Govt. of India for financial support. TA also thanks DST, New Delhi for providing funds (No. SR/ITS/02603/2008-2009) to present this paper in 2008 MRS Fall Meeting at Boston, USA. AG, IAW and SK thank UGC, Govt. of India, for a fellowship.

REFERENCES

1. T. Ahmad, R. Chopra, K. V. Ramanujachary, S. E. Lofland and A. K. Ganguli, J. Nanoscience Nanotechnology, **5**, 1840, 2005.
2. T. Ahmad, S. Vaidya, N. Sarkar, S. Ghosh and A. K. Ganguli, Nanotechnology, **17**, 1236, 2006.
3. A. K. Ganguli and T. Ahmad, J. Nanoscience Nanotechnology, **7**, 2029, 2007.
4. F. Chen, G. Q. Xu, T. S. A. Hor, Mater. Lett., **57**, 3282, 2003.
5. P. Barnickel, A. Wokuan, W. Sager, H. F. Eickel, J. Colloid Interface Sci., **148(1)**, 80, 1992.
6. M. L. Wu, D. H. Chen, T. C. Huang, Langmuir, **17**, 3877, 2001.
7. M. Boutonnet, J. Kizling, P. Stenius, Colloid Surf. **5(3)**, 209, 1982.
8. X. Zhang, K. Y. Chan, Chem. Mater., **15**, 451, 2003.
9. A. Agostaino, M. Catalano, M. L. Curri, M. D. Monica, L. Manna, L. Vasanelli, Micron, **31**, 253, 2000.
10. J. Easto, M. J. Hollamby, L. Hudson, Advance Colloid Interface Sci., **128**, 5, 2006.
11. A. M. Morales and C. M. Lieber, Science **279**, 208, 1998.
12. X. T. Zhou, N. Wang, H. L. Lai, H. Y. Peng, I. Bello, N. B. Wong and C. S. Lee, Appl. Phys. Lett. **74**, 3942, 1999.

13. T. J. Trentler, K. M. Hickman, S. C. Geol, A. M. Viano, P. C. Gibbans and W. E. Buhro, Science **270**, 1791, 1995.
14. H. Dai, E. W. Wong, Y. Z. Yu, S. S. Fan and C. M. Lieber, Nature **375**, 769, 1999.
15. W. Q. Han, S. S. Fan, Q. Q. Li and Y. D. Hu, Science **277**, 1287, 1997.
16. E. Leobandung, L. Guo, Y. Wang and S. Y. Chou, Appl. Phys. Lett. **67**, 938, 1995.
17. T. Ono, H. Saitoh and M. Esashi, Appl. Phys. Lett. **70**, 1852, 1997.
18. Z. W. Pan, Z. R. Dai and Z. L. Wang, Science **291**, 1947, 2001.
19. Z. W. Pan, Z. R. Dai, C. Ma and Z. L. Wang, J. Am. Chem. Soc. **124**, 1817, 2002.
20. Y. C. Kong, D. P. Yu, B. Zhang, W. Fang and S. Q. Feng, Appl. Phys. Lett. **78**, 407, 2001.
21. M. H. Huang, Y. Wu, H. Feick, N. Tran, E. Weber and P. Yang, Adv. Mater. **13**, 113, 2001.
22. J. J. Wu and S. C. Liu, J. Phys. Chem. B **106**, 9546, 2002.
23. J. J. Wu and S. C. Liu, Adv. Mater. **14**, 215, 2002.
24. X. M. Sun, X. Chen, Z. X. Deng and Y. D. Li, Mater. Chem. Phys. b, 99, 2002.
25. L. Vayssieres, Adv. Mater. **15**, 464, 2003.
26. C. H. Hung and W. F. Whang, Mater. Chem. Phys. **82**, 705, 2003.
27. C. Xu, G. Xu, Y. Liu and G. Wang, Solid State Commun. **122**, 175, 2002.
28. C. Xu, Y. Liu, G. Xu and G. Wang, Mater. Res. Bull. **37**, 2365, 2002.
29. C. Xu, G. Xu and G. Wang, J. Mater. Sci. **38**, 779, 2003.
30. R. S. Niranjan, Y. K. Hwang, D. K. Kim, S. H. Jhung, J. S. Chang, I. S. Mulla, Mater. Chem. Phys. **92**, 384, 2005.
31. S. Schiller, U. Heisig, K. Goedicke, H. Bilz, K. Steinfelder, Thin Solid Films **92**, 81, 1982.
32. S. C. Lee, J. H. Lee, T. S. Oh, Y. H. Kim, Sol. Energy Mater. Sol. Cells **75**, 481, 2003.
33. T. Stergiopoulos, I. M. Arabatzis, H. Cachet, P. Falaras, J. Photochem. Photobiol. A Chem **155**, 481, 2003.
34. D. Aurbach, A. Nimberger, B. Markovasky, E. Levi, E. Sominsky, A. Gedanken, Chem. Mater. **14**, 4155, 2002.
35. M. Miyauchi, A. Nikajima, T. Watanabe, K Hasimoto, Chem. Mater. **14**, 2812, 2002.
36. S. Y. Song, J. F. Ma, J. Yang, M. H. Cao and K. C. Li, Inorg. Chem. **44**, 2140, 2005.
37. A. Michalowicz, J. J. Girerd and J. Goulon, Inorg. Chem. **18**, 3004, 1979.
38. L. S. Birks and H. Friedman, J. Appl. Phys. **17**, 687, 1946.
39. B. E. Warren and B. L. Averbach, J. Appl. Phys. **21**, 595, 1950.
40. B. E. Warren and B. L. Averbach, J. Appl. Phys. **23**, 497, 1952.
41. O. G. Martinez, R. M. Rojas, E. Vila and J. L. M. de Vidales, Solid State Ionics **63**, 442, 1993.
42. R. R. Reeber, J. Appl. Phys. **41**, 5063, 1970.
43. Y. W. Duan, X. L. Kou and J. G. Li, Physica B **355**, 250, 2005.
44. R. Banerjee, E. A. Sperling, G. B. Thompson, H. L. Fraser, S. Bose and P. Ayyub, Appl. Phys. Lett. **82**, 4250, 2003.
45. T. Ahmad, K. V. Ramanujachary, S. E. Lofland and A. K. Ganguli, J. Mater. Chem. **14**, 3406, 2004.

Nanowires: Synthesis and Characterization

Mater. Res. Soc. Symp. Proc. Vol. 1142 © 2009 Materials Research Society 1142-JJ09-04

Optical and magnetic properties of Co nanostructure thin films

D. Schmidt, T. Hofmann, A. C. Kjerstad, M. Schubert, and E. Schubert
Department of Electrical Engineering and Nebraska Center for Materials and Nanoscience,
University of Nebraska-Lincoln, Lincoln, NE, 68588-0511, U.S.A.

ABSTRACT

We report on optical and magnetic properties of two substantially different Cobalt nanostructure thin films deposited at an oblique angle of incidence of 85° away from the substrate normal. A columnar thin film grown without substrate rotation and a nanocoil sculptured thin film produced by glancing angle deposition with substrate rotation are compared. Generalized spectroscopic ellipsometry determines the optical constants of the films in the spectral range from 400 nm to 1000 nm, and the magnetic properties are analyzed with a superconducting quantum interference device magnetometer. Both nanostructure thin films show highly anisotropic optical properties such as strong form birefringence and large dichroism. Co slanted columnar thin films are found to be monoclinic. Magnetic measurements show hysteresis anisotropy and extremely large coercive fields at room-temperature with respect to a magnetic field either parallel or perpendicular to the nanostructures long axis. We find coercive fields of 3 kOe for our nanostructures.

INTRODUCTION

Glancing angle deposition (GLAD) is a physical vapor deposition process (evaporation or sputtering) with particle flux at very oblique angle of incidence (typically > 80°) onto the substrate [1]. A concurrent growth mechanism due to geometrical self-shadowing in combination with limited adatom mobility governs the deposition process. With a controlled substrate motion GLAD allows for a self-organized bottom-up fabrication of three dimensional nanostructures ranging from slanted or straight columns to screws and helices. A broad range of materials such as semiconductors, dielectrics, or metals can be tailored with this technique on practically all surfaces. Due to their small size and shape anisotropy such columnar and sculptured thin films may possess intriguing optical, mechanical, electrical and magnetic properties [2].

In this paper we study structural, optical, and magnetic properties of GLAD grown Co nanostructures with a strong focus on generalized ellipsometry (GE) and superconducting quantum interference device magnetometer (SQUID) investigations. We show that GE analysis is not only limited to accurately determine principal optical constants and intrinsic birefringence but also has excellent capabilities in identifying geometrical parameters such as orientation of nanostructures.

Few reports have been published already on GLAD growth of randomly and also periodic arranged nanostructures with emphasis on magnetic hysteresis investigations [3-5]. However, with decreasing geometry in the nm range also mechanical and optical properties differ drastically from their bulk material and, for instance, form-induced polarization confinement and quantization effects need to be taken into account. Understanding these properties is of high importance for tailoring new artificial materials. Non-destructive optical techniques such as GE have proven to be extremely suitable for determining structural and physical properties (i.e. dielectric tensor) of highly anisotropic thin films. It has been recently reported for orthorhombic

[6] and triclinic [7] thin films as well as monoclinic slanted GLAD columns from Ti and Cr [8,9].

EXPERIMENT

Cobalt nanostructures were deposited at room temperature by electron-beam GLAD in a customized ultra high vacuum chamber onto silicon substrates. The [001] p-type Si substrates had a native oxide layer of ≈ 3 nm. The distance between source and the (x-y-z) sample manipulator is 460 mm. The deposition angle measured between the incident particle flux direction and the substrate normal was set to 85°. Figure 1 depicts a high-resolution field-emission scanning electron microscope (SEM) image of an edge view of each sample. The slanted columnar thin film (Figure 1A) was deposited without substrate rotation whereas the substrate was rotated counterclockwise at 0.25 rpm while growing the chiral sculptured thin film (nanocoils) shown in Figure 1B. Angle-resolved (angle of incidence Φ_A and in-plane rotation angle φ) spectroscopic Mueller matrix ellipsometry measurements were performed using a commercial instrument (M2000^{TM}, J. A. Woollam Co., Inc.) within the spectral range from $\lambda = 400$ nm to 1000 nm. The ellipsometer was mounted on an automatic variable Φ_A and sample rotator φ stage. Φ_A was varied from 45° to 75° in steps of 10°, while φ was varied from 0° to 360° in steps of 5°. The polarizer-compensator-sample-analyzer ellipsometer is capable of measuring 11 out of 16 Mueller matrix elements normalized to M_{11} (except for elements in fourth row) [10].

The obtained spectra for the slanted columnar thin film were modeled with a single biaxial (triclinic) layer. Euler angles (φ, θ, ψ) that transform the Cartesian coordinate system into the sample coordinates represent the orientation of the nanostructure. Similarly, the nanocoils where described by a graded biaxial (orthorhombic) layer what allows for rotation of Euler angle φ (in plane orientation) to account for the substrate rotation during the growth process. The major optical constants presented in Figure 3 were extracted on a point-by-point basis; i.e., without any physical lineshape implementations. A detailed description of this method can be found in [11] and references therein.

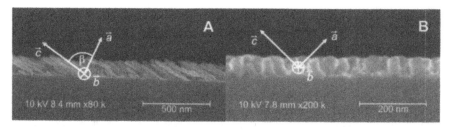

Figure 1 SEM micrograph of slanted Co nanocolumns (A) and Co nanocoils (B). The overlaid coordinate system indicates the orientation of the biaxial system with its internal c-axis along the nanocolumns/nanocoils and b-axis parallel to the film interface. In (A) the monoclinic angle β between axes a and c is 80.6° whereas in (B) the system is orthorhombic and rotates with respect to the surface normal along the nanocoils.

Magnetization properties (hysteresis loops) were recorded with a SQUID (MPMS XL7, Quantum Design) at room temperature. Measurements parallel and perpendicular to the long axis of the structures were taken to examine differences in hysteresis loops due to sample anisotropy.

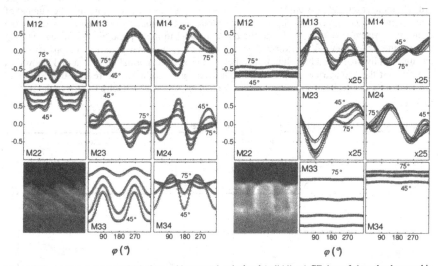

Figure 2 Exemplary experimental (circles) and best-match calculated (solid lines) GE data of slanted columnar thin film (left) and nanocoils (right) versus sample azimuth angle φ and angle of incidence Φ_A at $\lambda = 850$ nm. Numbers in the right panel correspond to the magnification of the data shown. Note that the chiral nanostructures do not reveal an isotropy orientation; i.e., a sample position φ at which all off-diagonal Mueller matrix data (M_{13}, M_{14}, M_{23}, M_{24}) vanish, but which is the case near $\varphi = 0°$ and $\varphi = 180°$ for the nanocolumns.

DISCUSSION

Optical and structural properties

Figure 2 depicts selected Mueller matrix data (GE) obtained from both Co nanostructures, shown in Figure 1, at an exemplary wavelength of $\lambda = 850$ nm. Mueller matrix elements not shown in the figure can be obtained by symmetry as described earlier [12]. Note that model and experimental data are in perfect agreement in both cases for all wavelengths in the investigated spectral region from $\lambda = 400$ nm to 1000 nm (data omitted for brevity here). The off-diagonal Mueller matrix data (M_{13}, M_{14}, M_{23}, M_{24}) exhibit the highly anisotropic nature of the Co nanostructures. These elements are zero for all angles of incidences Φ_A at all wavelengths for isotropic samples. For the slanted columnar Co thin film pseudo-isotropic sample orientations can be identified at $\varphi \approx 0°$ and $\varphi \approx 180°$, which coincide with orientations of the sample when the direction of the nanocolumns is parallel to the plane of incidence. Such orientations are missing in the case of the nanocoil sample. We further obtained that the slanted Co columnar thin film possesses monoclinic optical properties with an angle $\beta = 80.6°$ (Figure 1A), and which is in agreement with our previous reports on Ti and Cr slanted columnar thin films [8, 9]. This monoclinic angle can be understood as a structural property of slanted columnar thin films. At the bottom of the structure, charge exchange is possible due to their conducting nucleation layer whereas this is not possible at the isolated top of the columns. Therefore, the overall dipole moment for electric fields perpendicular to the columns and within the slanting plane is tilted towards the surface normal. For the chiral nanocoils our model calculations show that the orthorhombic coordinate system is rotating with the c-axis along the windings and the number of turns is in perfect agreement with the actual number of rotations during growth. Hence, the coordinate system is not oriented with the c-axis parallel to the surface normal (along the long axis of the nanostructure) as expected in the case of an uniaxial material composed of, for example, individual straight columns parallel to the surface normal [13]. Table 1 shows an overview of the geometrical properties of both investigated samples and reveals that SEM and GE values are in excellent agreement. Inclination angles are measured with respect to the surface normal.

Table 1 Overview of the geometrical properties of our Co samples obtained by SEM and GE.

	Slanted Columns		Nanocoils	
	SEM	GE	SEM	GE
Thickness	125 nm	113.6 nm	63 nm	63.7 nm
Inclination θ	55°	55.3°	---	37.5°

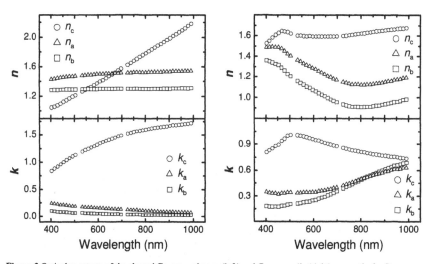

Figure 3 Optical constants of the slanted Co nanocolumns (left) and Co nanocoils (right), respectively. Strong birefringence and dichroism can be observed in the optical constants of both nanostructures. Note that wavelengths around 520 nm and 720 nm are disregarded due to detector noise.

Figure 3 depicts wavelength dependencies of refractive indices n_i and absorption coefficients k_i ($i = a, b, c$) that differ drastically from bulk material [14]. Strong birefringence and dichroism can be observed in the investigated spectral region between all polarizabilities. The index of refraction n_c along the slanted nanocolumns c-axis is crossing both other refractive indices n_b and n_a. This observation is in not in full agreement with the order of principal refractive indices for similar biaxial mediums ($n_c > n_b > n_a$) reported earlier but true for wavelength greater than $\lambda = 670$ nm [13]. Such intersections are not present in case of the nanocoils with respect to n. However, here the absorption coefficients k_a and k_b are crossing at around $\lambda = 800$ nm. In general, n_c and k_c of the slanted nanocolumns have a strong wavelength dependence in contrast to the optical constants along the a- and b-axes. Note that there is almost no absorption parallel to the film interface. Considering the nanocoils, both refractive indices n_i and absorption coefficients k_i have moderate wavelength dispersion along all axes. It can be seen that within the investigated spectral region both parameters n_c and k_c are largest.

In general, the optical constants of both nanostructures are significantly different even though they are from the same material and geometries are in the same order of magnitude. The intrinsic coordinate system of the nanocoil thin film (Figure 1B) exhibits no monoclinic angle in contrast to the slanted columnar thin film. Hence, we expect the internal structure of the slanted columns to be different from the nanocoils. This assumption is subject to further investigations.

Magnetic analysis

Magnetic hysteresis loops where carried out at room temperature using a SQUID magnetometer (Figure 4). In both cases the external B field was applied parallel and perpendicular to the long axis of the nanostructures. In comparison with an e-beam evaporated bulk-like sample (600 nm, normal incidence), which exhibits a coercive field $H_c \approx 30$ Oe, both Co nanostructures show greatly enlarged coercivities due to their small diameter and strong shape anisotropy. The easy axis is perpendicular to long axis of the nanostructures. A coercive field $H_c = 3$ kOe for the slanted columnar thin film is comparable with what was reported for electrodeposited Co nanowires with similar diameter [15].

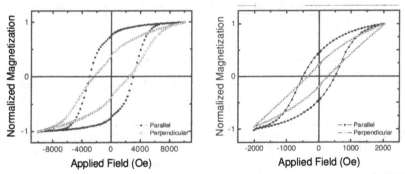

Figure 4 Room temperature magnetic hysteresis loops of slanted Co nanocolumns (left) and Co nanocoils (right). A strong coercive field $H_c = 3$ kOe is observed for the slanted nanocolumns.

CONCLUSIONS

We studied structural, optical, and magnetic properties of electron beam GLAD Co nanostructures. We obtained a complete set of optical constants for biaxial Co slanted nanocolumns and nanocoils grown at a glancing angle of 85°. With respect to the nanocolumns dielectric constants (n_a, n_b, n_c) and absorption coefficients (k_a, k_b, k_c) are determined along the principal axis of a monoclinic lattice and the angle β was found to be 80.6°. Structural information derived from best-model calculations are in both cases in very good agreement with SEM investigations. Furthermore, we have shown that nanocoil thin films and slanted columnar thin films made from Co show hysteresis anisotropy parallel and perpendicular to the structure. The latter one exhibits a large coercive field of H_c = 3 kOe.

ACKNOWLEDGMENTS

The authors acknowledge financial support from NSF in SGER and MRSEC QSPIN, Coe, UNL, and J.A. Woollam Foundation.

REFERENCES

[1]   K. Robbie, M. J. Brett, J. Vac. Sci. Technol. A **15**, 1460 (1997)
[2]   A. Lakthakia and R. Messier, Sculptured Thin Films (SPIE Press, Bellingham, 2004).
[3]   B. Dick, M. J. Brett, T. J. Smy, M. R. Freeman, M. Malac, and R. F. Egerton, J. Vac. Sci. Technol A **18**, 1838 (2000).
[4]   F. Liu, C. Yu, L. Shen, J. Barnard, and G. J. Mankey, IEEE T. Magn. **36**, 2939 (2000).
[5]   A. Lisfi and J. C. Lodder, Phys. Rev. B **63**, 174441 (2001).
[6]   M. Schubert and W. Dollase, Opt. Lett. **27**, 2073 (2002).
[7]   M. Dressel, B. Gompf, D. Faltermeier, A. K. Tripathi, J. Pflaum, and M. Schubert, Opt. Express **16**, 19770 (2008).
[8]   D. Schmidt, B. Booso, T. Hofmann, A.Sarangan, E. Schubert, and M. Schubert, Appl. Phys. Lett. **94** (2009) (in press).
[9]   D. Schmidt, B. Booso, T. Hofmann, A.Sarangan, E. Schubert, and M. Schubert, Opt. Lett. (2009) (in submission).
[10] H. G. Tompkins and E. A. Irene, eds., Handbook of Ellipsometry (Springer, Heidelberg, 2004).
[11] M. Schubert, Ann. Phys. (Leipzig) **15**, 480 (2006).
[12] D. Schmidt, E. Schubert and M. Schubert, phys. stat. sol. (a) **205**, 748 (2008).
[13] I. J. Hodgkinson and Q. H. Wu, Birefringent Thin Films and Polarizing Elements (World Scientific, Singapore, 1998).
[14] E. D. Palik, ed., Handbook of Optical Constants of Solids (Academic Press, Boston, 1991).
[15] Y. Henry, K. Ounadjela, L. Piraux, S. Dubois, J.-M. George, and J.-L. Duvail, Eur. Phys. J. B **20**, 35 (2001).

Carbon Nanostructure Processing

Mater. Res. Soc. Symp. Proc. Vol. 1142 © 2009 Materials Research Society   1142-JJ10-01

Surface Oxidation of Single-Walled Carbon Nanotubes with Oxygen Atoms

Luciana Oliveira, Thomas Debies[1] and Gerald A. Takacs[*]

Department of Chemistry, Center for Materials Science and Engineering,
Rochester Institute of Technology, Rochester, NY, 14623, U.S.A.
[1]Xerox Corporation, Webster, NY, 14580, U.S.A.

ABSTRACT

Single-walled carbon nanotube (SWNT) powder was surface oxidized with gaseous oxygen atoms produced by low-pressure: (1) vacuum UV (VUV) (λ = 104.8 and 106.7 nm) photo-oxidation and (2) microwave (MW) plasma discharge of an Ar-O_2 mixture. X-ray photoelectron spectroscopy (XPS) was used to detect the carbon- and oxygen-containing functional groups in the top 2-5 nm of the sample's surface. VUV photo-oxidation showed more of the C-O-C and O-C=O functional groups compared to the carbonyl, C=O, moiety. The results of this study are compared to previous investigations where ozone produced from high-pressure, UV photo-oxidation (λ = 184.9 and 253.7 nm) was mainly reacted with SWNT powder and paper. The presence of UV or VUV radiation shows higher levels of oxidation than the MW results which were conducted in the absence of radiation from the plasma.

INTRODUCTION

Gas-phase oxidation of carbon nanotubes (CNTs), which would eliminate the liquid waste generated from solution-phase studies, could be a valuable dry technique for modifying the top layers of the surface, thus helping with the manufacture of nanoelectronic devices [1-3].

In this study, low-pressure, gas-phase vacuum UV (VUV) photo-oxidation of single-walled carbon nanotube (SWNT) powder is studied at room temperature with wavelengths from excited Ar atoms (λ = 104.8 and 106.7 nm) that have sufficient energy to photo-dissociate gaseous oxygen and result in chemical modification of the surface. In addition, the reaction of oxygen atoms with SWNT powder was carried out in the absence of VUV radiation using a microwave discharge (MW) of an Ar and O_2 mixture. The results of this study are compared to our previous experiments using atmospheric pressure, UV photo-oxidation (184.9 and 253.7 nm) of SWNT powder [1] and paper [2] where ozone is the principal reactant with the nanotubes.

EXPERIMENTAL

The SWNT powder, which was purchased from Strem Chemicals, Inc., Newburyport, MA consists of diameters from 0.7 to 2 nm, lengths from 2-20 μm and usually occur in bundles of 20 tubes. The powder was used as received and placed within a well formed in a quartz block during both the oxidation and XPS analysis.

Low-pressure argon MW plasmas, operating at a frequency of 2.45 GHz and absorbed power of 30 – 40 W (the difference between the forward and reflected power), and 25 – 31 W were used as the source of VUV radiation [3] and oxygen atoms, respectively, to modify the surface of SWNTs located downstream from the plasma. For VUV photo-oxidation, the samples were placed 23.8 cm downstream from the MW discharge of Ar. Oxygen was introduced into the vacuum system about 3 cm above the sample. The argon and oxygen flow rates were 50 and 10 sccm, respectively. The reaction chamber pressure was maintained at (4.3 - 4.8) x 10^1 Pa. At the reaction time associated with this distance, charged particles and metastables from the plasma contribute only negligibly because of recombination and deactivation processes occurring in transit to the sample [3].

For the MW discharge of Ar-O_2 mixtures, the construction and operation of the discharge flow system was similar to that used in the study of the gas-phase O + HBr reaction [4]. Atomic oxygen was generated from an Ar-O_2 mixture having flow rates of 50 and 10 sccm, respectively, with the pressure in the chamber at (1.3 – 4.0) Pa. The vacuum system was designed so that the discharge was located ca. 41 cm upstream from the SWNTs and the radiation from the discharge was not directed at the sample.

The samples were analyzed with a Physical Electronics Model 5800 XPS that examines the top 2 – 5 nm of a sample's surface using a take-off angle of 45° between the sample and analyzer. A region of about 800 μm in diameter was analyzed. The quartz block was mounted directly in an XPS sample holder. The monochromatic Al K_α (1486 eV) x-ray beam irradiated the well of the quartz block and the electron optics of the analyzer were focused to accept only photoelectrons emitted from the nanotubes. The quantitative analyses are precise to within 5% relative for major constituents and 10% relative for minor constituents. To minimize radiation damage, the samples were charge neutralized with a flood of low energy electrons from a BaO field emission charge neutralizer.

RESULTS and DISCUSSION

Quantitative XPS analysis of four untreated samples of SWNT powder showed only the presence of carbon and oxygen with an atomic percentage (at%) for oxygen of 4.8 ± 0.4 which is in good agreement with our previously reported control of 4.9 [1]. Occasionally, after treatment, small amounts of Co, N and Si were detected.

Figure 1 compares the O at% results as a function of treatment time for VUV photo-oxidation of SWNT powder (this work) with that previously reported for the UV photo-oxidation of SWNT powder [1] and SWNT paper [2]. Surface oxidation rapidly occurs to reach a common level of saturation using the two different treatment methods and two different forms of SWNTs.

The overlapped C_{1s} spectra for a control and for SWNT powder samples treated for 1 h with VUV (this work) and UV [1] photo-oxidation are presented in Figure 2.

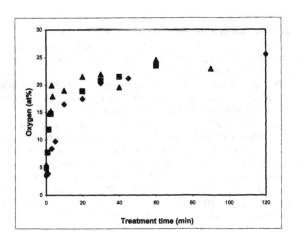

Figure 1. Plot of atomic percent of oxygen as a function of treatment time for SWNT powder treated with VUV (▲) [this work] and UV (■) photo-oxidation [1], and SWNT paper treated with UV (♦) photo-oxidation [2].

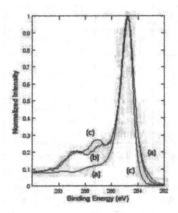

Figure 2. C_{1s} XPS spectra for SWNT powder: (a) untreated, and treated for (b) 1 h with UV photo-oxidation [1] and (c) 1h with VUV photo-oxidation.

The principal peak due to carbon-carbon bonding at 284.8 eV dominates the spectra in Figure 2, but complex spectral features due to carbon-oxygen bonding are evident at higher binding energies. Curve fitting was completed for the C_{1s} spectra.

Binding energy (B. E.) values reported in the literature [5] were utilized to assign the following peaks with increasing binding energy: C-C sp^2, C-C sp^3, C-O-C as ether and/or epoxy, C=O, O-C=O, O=C-O-C=O and/or O-(C=O)-O. The C_{1s} binding energy for the anhydride group, O=C-O-C=O [5], has been reported to have similar values as the carbonate-like, O-(C=O)-O, moiety [6]. As a result of the variety of species and similar contributions, the high binding energy region of the spectra has a broad undulating appearance. The peaks were modeled with different full width at half maximum by assignment. The C_{1s} peaks due to carbon-carbon bonding were fit with peaks whose full width at half maximum was about 0.7 eV while the peak due to energy loss was fit with a peak with a full width at half maximum of 1.8 eV. The percentage of carbon species was estimated by curve fitting the minimum number of peaks necessary to achieve chi-squared values of 2.0 or less. The binding energies, peak assignments and absolute percentages of carbon are reported in Table I. The absolute percentages of carbon were calculated by multiplying the C at% from the quantitative analyses times the percent contribution from the species obtained from the curve fitting. Therefore, the sum of the carbon concentrations in Table I equal the concentration of carbon obtained from the quantitative analyses and not 100%.

Table I. Results of Absolute % of C-containing Groups for SWNT Powder Treated for 1 h with VUV and UV Photo-oxidation, and 1 h with Oxygen Atoms Produced from a MW Discharge of an Ar-O_2 Mixture

Assignment	B. E. (eV)	Untreated	VUV	UV [1]	MW
C-C sp^2	284.7	48.2	29.5	28.1	37.7
C-C sp^3	285.1	28.4	15.5	19.8	20.5
C-O-C, C—C (O)	286.0	3.8	9.6	6.8	5.7
C=O	287.0	4.7	4.4	9.1	8.2
O-C=O	288.6	3.8	8.8	6.8	4.1
O=C-O-C=O, O-(C=O)-O	289.8	2.8	3.7	3.8	3.3
Energy Loss	292.0	2.8	2.2	1.5	2.5
Total =		94.5	73.7	75.9	82.0

Figure 3 shows that the SWNTs treated downstream from the MW Ar/O_2 mixture in the absence of photons contained lower levels of oxidized species (ca. 15 at% O) than the nanotubes exposed to VUV and UV (Fig. 1 [1]) photo-oxidation (ca. 24 at% O). The level of oxidation obtained by the MW discharge of a 5:1 mixture of Ar:O_2 are in good agreement with those reported when carbon nano-fibers/tubes are treated with 27.12 MHz RF discharges containing 1:1 mixtures of Ar:O_2 [7, 8] and pure O_2 [9]. The shapes of the C_{1s} spectra for the MW discharge experiments (not shown here), are similar to those reported in Figure 2.

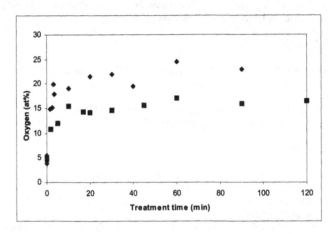

Figure 3. Plot of atomic percent of oxygen as a function of treatment time for SWNT powder samples treated with VUV photo-oxidation (♦) and the MW discharge method for producing oxygen atoms (■).

Table I also reports the absolute carbon percentages for the MW Ar/O_2 experiments compared to the VUV and UV photo-oxidation techniques. In contrast to the MW and UV experiments, the presence of VUV photons (11.6 and 11.8 eV), which have sufficient energy to readily break carbon-carbon bonds and possibly form excited oxygen atoms, produces more of the C-O-C and O-C=O functional groups than the C=O, moiety.

The O_{1s} spectra for the VUV, MW and UV [1] gas-phase oxidation methods (not shown here) showed broad peaks with no energy resolved structure.

The reaction of UV-produced ozone with SWNT powder [1] resulted in a similar level of oxidation as the VUV photo-oxidation method (ca. 24 at% O, Figure 1). For MWNT paper, the saturation level resulting from UV photo-oxidation has been reported to be ca. 2.8 at% O [1] compared to ca. 7.4 at% O for VUV photo-oxidation [3]. The greater curvature (i.e., strain) on the smaller diameter SWNTs probably accounts for the increased reactivity relative to the larger diameter outer shells of the MWNTs.

CONCLUSIONS

SWNT powder was surface oxidized by O atoms produced by: (1) photo-oxidation in the presence of VUV radiation (λ = 104.8 and 106.7 nm) and (2) MW plasma discharge of an Ar-O_2 mixture in the absence of radiation from the discharge. The MW experiments achieved lower saturated levels of oxidation (ca. 15 at% O) than with VUV photo-oxidation (ca. 24 at% O). The presence of the highly energetic VUV photons and possibly excited oxygen atoms resulted in the production of more C-O-C and O-C=O functional groups than the carbonyl, C=O, moiety. The increased reactivity of O atoms with SWNTs is probably due to the greater curvature (i.e., strain) on the smaller diameter tubes compared to the larger diameter outer shells of MWNTs.

Acknowledgements

The authors gratefully acknowledge the help of Dr. K.S.V. Santhanam for the supply of SWNTs and his support and encouragement during the research.

*To whom correspondence should be addressed. Phone 585-475-2047. Fax: 585-475-7800. E-mail address: gatsch@rit.edu.

REFERENCES

1. M. Krysak, A. Jayasekar, B. Parekh, L. Oliveira, T. Debies, K. S. V. Santhanam, R. A. DiLeo, B. J. Landi, R. P. Raffaelle and G. A. Takacs in: *Polymer Surface Modification: Relevance to Adhesion*, K. L. Mittal (Ed.), VSP/Brill, Leiden, in press (2008).
2. B. Parekh, T. Debies, C. M. Evans, B. J. Landi, R. P. Raffaelle and G. A. Takacs, *Mater. Res. Soc. Symp. Proc.* **887**, 3-8 (2006).
3. M. Krysak, B. Parekh, T. Debies, R. A. DiLeo, B. J. Landi, R. P. Raffaelle and G. A. Takacs, *J. Adhesion Sci. Technol.* **21**, 999-1007 (2007).
4. G. A. Takacs and G. P. Glass, *J. Phys. Chem.* **77**, 1182 (1973).
5. G. Beamson and D. Briggs, *High Resolution XPS of Organic Polymers: The Scienta ESCA300 Database*, Wiley, Chichester (1992).
6. J. F. Moulder, W. F. Stickle, P. E. Sobol and K. D. Bomben, in: *Handbook of X-ray Photoelectron Spectroscopy*, J. Chastin and R. C. King Jr. (Eds), p. 216, Physical Electronics, Eden Prairie, MN (1995).
7. H. Bubert, S. Haiber, W. Brandl, G. Marginean, M. Heintze and V. Bruser, *Diamond Related Mater.* **12**, 811-815 (2003).
8. M. Heintze, V. Bruser, W. Brandl, G. Marginean, H. Bubert and S. Haiber, *Surf. Coat. Technol.* **174-175**, 831-834 (2003).
9. V. Bruser, M. Heintze, W. Brandl, G. Marginean and H. Bubert, *Diamond Related Mater.* **13**, 1177-1181 (2004).

Mater. Res. Soc. Symp. Proc. Vol. 1142 © 2009 Materials Research Society   1142-JJ10-05

Optical properties of nanoparticle-doped azobenzene liquid crystals

R. M. Osgood III[1], D. M. Steeves[1], L. E. Belton[1], J. R. Welch[1], R. Nagarajan[1], Caitlin Quigley[1], G. F. Walsh[1], N. V. Tabiryan[2], S. Serak[2], B. R. Kimball[1]

[1]US Army Natick Soldier Research, Engineering, and Development Center, Natick, MA 01760 USA
[2] Beam Engineering for Advanced Measurements Co., 809 So. Orlando Ave., Suite I, Winter Park, FL, 32789, USA

ABSTRACT

Liquid crystals (LCs) are anisotropic fluids exhibiting orientational order that can be controlled with external fields. The optical nonlinearity of LCs can be enhanced by doping with azobenzene dyes or nanoparticles such as carbon nanotubes (CNTs). Previous studies have demonstrated modifications of the LC's nematic-isotropic phase transition temperature by the addition of CNTs. In the present paper, we disperse single-wall and multi-wall CNTs in LC cells at the level of 0.1% by weight. We find that the addition of CNTs enhances the response speed of the azobenzene LC. We have developed a theory that allows quantitative characterization of the nonlinear transmission process of a system of crossed polarizers comprising the LC cell.

INTRODUCTION

Liquid crystals (LCs) are a condensed phase with intermediate (orientational, not positional) order in between that of a crystalline solid and a random liquid; the director (the axis of the average LC's molecular orientation) of the anisotropic nematic LC (NLC) phase can be controlled with electromagnetic fields. The nematic phase undergoes a phase transition and melts into an isotropic phase at the so-called clearing temperature.

LCs have many important applications in electro-optics and photonics. The transmission of crossed polarizers comprising a LC cell can be switched on or off with an applied voltage within milliseconds. All-optical processes can take place at much shorter time scales. One class of candidate LC materials for rapid response to illumination is azobenzene LCs (azo LCs). Azobenzene molecules photoisomerize from the trans state to the cis state on the picosecond time scale, and may enable ultrafast (e.g. pulsed laser) optical switching at low illumination powers. Nonlinear optical processes in these materials may take place at microwatt and even nanowatt power levels under cw beams [1-3] and with ~ 1 mJ laser pulses [4]. Here the accumulation of cis molecules drives the nematic-to-isotropic phase transition.

There have been a few attempts reported in the literature to enhance the optical nonlinearity of LCs by adding nanoparticles [5-9]. In Ref. 10, the melting temperature of the LC component was increased by the incorporation of anisotropic multi-wall CNTs (MWCNTs) within a small composition range, while outside this range the MWCNTs did not influence the melting temperature; isotropic fillers did not change the melting temperature.

In this paper, we report on attempts to reduce the threshold energy levels required for the photoinduced nematic-isotropic phase transition by adding conductive nanoparticles to azo LCs. We were interested in the antenna-like behavior of conductive nanoparticles (MWCNTs and single-wall CNTs or SWCNTs); we believed that such "nanoantennas" (similar to those reported in Ref. 11) might intensify the local electromagnetic field and reduce the threshold for the photoinduced phase transition, due to the large difference between the dipole moments of trans and cis isomers. When one illuminates an azo LC with blue-green wavelengths, we expect trans to cis conversion, possibly with an accompanying melting transition. The transmission of a twist or planar cell between crossed polarizers decreases dramatically when enough trans molecules convert to cis. We expected that adding nanoparticles (carbon nanotubes) would increase the amount of cis molecules present, and therefore decrease the response time of the nanoparticle-doped LC cell.

EXPERIMENTAL

We used two developmental dyes, entitled "Dye-57" (a mixture of 85% 5CB and 15% of a mixture of CPND-5 and CPND-7), and "Dye-68" (a mixture of 68% LC1005 and 32% of a mixture of CPND-6 and CPND-8). The CPND dyes were reported in Ref. [12]. LC 1005 is a room temperature azo NLC [13]; its nonlinear optical properties were studied in Refs. 2-4. Dye-57 and Dye-68 were often, but not always, diluted in 5CB by a ratio of 5:1 in volume, in order to work with thicker cells (which are easier to construct).

LC twist cells were fabricated by rubbing polyvinyl alcohol (PVA), which had been deposited onto glass substrates and then annealed at 120°C for 20 minutes. Cell thickness was typically 1 μm, but ranged from 0.5 to 5 μm, depending on the spacers and dye used. Two surfaces were oriented 90° to one another and epoxied together into a "twist" cell, which rotates the incident polarization by 90°. Alternatively, a "planar" cell was made by orienting the two rubbed surfaces 180° to one another.

Nanoparticles were introduced into the samples in the following manner. SWCNTs (1 nm diameter, approximately 1 μm long, nominal 0.1 μΩ-cm resistivity), pre-dispersed in methanol, were sonicated into the LC. A separate batch of samples was prepared using powdered MWCNTs (140 nm in diameter and 3 microns long). The MWCNTs were sonicated in a heterocyclic aromatic tertiary amine (to reduce bundling effects), combined with the Dye-68 or Dye-57 (composition was about 0.1% by weight), and the solvent was removed by evaporation. The addition of carbon nanotubes increased optical density moderately in the blue-green part of the spectrum, as illustrated for a typical sample pair in Figure 2. Samples were made with both this direct dye concentration, and with a 1:5 volume dilution in 5CB LC (we denote the latter cells "Dye-57,-68/5CB").

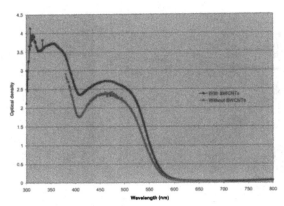

Figure 1: Optical density in the visible range of Dye-68/5CB twist cells with and without SWCNTs.

LC cells were placed between crossed polarizers, with the initial polarization parallel to the rubbing direction of the front glass surface for the twist cells (45° for the planar cells). We measured the transmission of a continuous wave (cw) Ar-Kr laser (power up to 120 mW, beam diameter 2 mm, calculated power density 3.8 W/cm^2) at 530 nm (shorter wavelengths experienced significant absorption) by opening a shutter, which triggered the oscilloscope (50 Ω coupling); we recorded the output voltage V on a fast photodiode detector $vs.$ time. We converted voltage to transmission normalizing to the measured transmission value at low power, thus obtaining a plot of transmission vs. time ("T vs. t"), where the shutter was opened at t = 0. For twist samples, unless the absorption was low, the sample converted entirely to cis and the transmission dropped to zero. Some planar samples exhibited an oscillating transmission vs. time, while others very quickly (tens of ms) converted entirely to cis. While the time scales under discussion here are not desirable for ultrafast switching, we anticipate that high peak power lasers, used in conjunction with the azo LC cells, will enable much faster switching.

In Figure 2 we display T vs. t for two Dye-68/5CB twist cells, one with MWCNTs and one without; in Figure 3, we display T vs. t for Dye-68/5CB twist cells with and without pre-dispersed SWCNTs. The addition of both types of CNTs dramatically reduced the response time (transmission drops to zero much faster than without the nanoparticles). The addition of MWCNTs increased absorption similar to the effect of SWCNTs illustrated in Fig. 1. CNTs did not reduce the response time in Dye-57/5CB samples as dramatically, perhaps because the absorption in Dye-57/5CB cells was considerably higher (~ 40% at 530 nm) than in Dye-68/5CB cells. Planar cells also did not show as dramatic an effect with introduction of MWCNTs or SWCNTs.

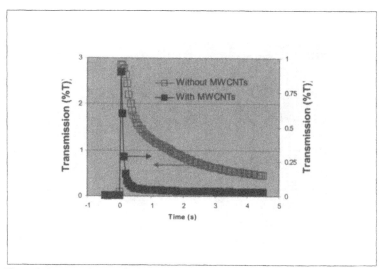

Figure 2: Measured transmission as a function of time for two Dye-68/5CB twist cells, with and without MWCNTs. Both are about 5 μm thick. Laser power was 120 mW.

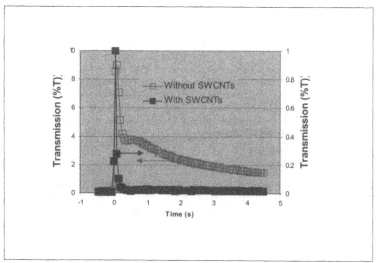

Figure 3: Measured T vs. t for two Dye-68/5CB twist cells, with and without SWCNTs. The sample without SWCNTs was about five microns thick. Laser power was 100 mW.

We also added SWCNTs to pure Dye-57 and Dye-68 dyes (no dilution with 5CB), and there was considerably less effect from the SWCNTs, as shown in Figure 4 (~2x increase in absorption, but little reduction in response time). Because of their greater absorption, the samples with undiluted dyes become opaque in much shorter times (tens of milliseconds) than the Dye-57/5CB and Dye-68/5CB cells.

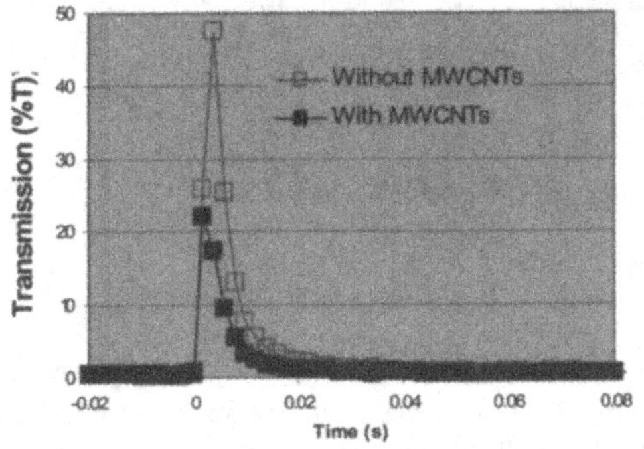

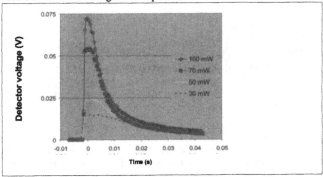

Figure 4: Measured T vs. t for two Dye-57 (undiluted) twist cells. The sample with MWCNTs is about 1 µm thick. Laser power was 100 mW.

As an illustration of the dependence of the transmission on laser beam power, we display in Figure 5 V vs. t for an approximately 1 µm thick undiluted planar Dye-57/MWCNT cell illuminated with four different laser powers. Trans to cis conversion occurs much faster at higher laser powers.

Figure 5: Measured V vs. t for a planar MWCNT-doped Dye-57 cell ~1 µm thick.

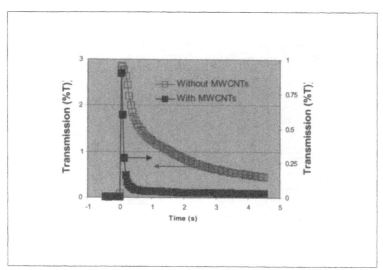

Figure 2: Measured transmission as a function of time for two Dye-68/5CB twist cells, with and without MWCNTs. Both are about 5 μm thick. Laser power was 120 mW.

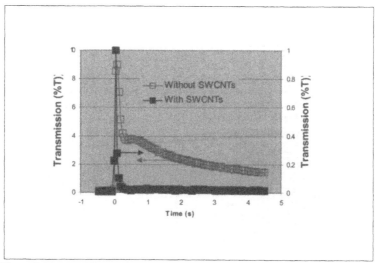

Figure 3: Measured T vs. t for two Dye-68/5CB twist cells, with and without SWCNTs. The sample without SWCNTs was about five microns thick. Laser power was 100 mW.

In our case, concentration of cis isomers is the driving factor and, at the vicinity of a phase transition, we can assume $A = a(N_c - N^*)$ with a being a positive constant. Here N^* is the critical concentration of cis isomers inducing phase transition into isotropic state. Nematic-isotropic phase transitions can be regarded as second order for any practical purpose, and we can ignore the term $\sim Q^3$. Thus minimizing the free energy (3) we obtain the following relationship:

$$Q = \sqrt{\frac{a(N^* - N_c)}{C}}$$

(4)

Expressing the proportionality of the optical anisotropy of the LC to the order parameter in the form $\Delta n = bQ$, we finally obtain

$$T = T_0 \sin^2\left(\frac{\pi L}{\lambda} \Delta n_0 \sqrt{1 - \frac{N_c}{N^*}}\right)$$

(5)

where $\Delta n_0 = b(aN^*/C)^{1/2}$ is the optical anisotropy of the LC before illumination ($N_c = 0$). Solving Eq. (1) and assuming initially zero cis concentration, we rewrite Eq. (5) as

$$T = T_0 \sin^2\left(\Delta\Phi_0 \sqrt{1 - r\left(1 - e^{-t/\tau}\right)}\right)$$

(6)

The three parameters that control the dynamics of the transmission are:

$\Delta\Phi_0 = \dfrac{\pi L}{\lambda} \Delta n_0$, the phase difference between ordinary and extraordinary waves, $r = \dfrac{N_s}{N^*}$, the ratio of the cis isomers obtained at the steady state N_s to the critical concentration, and the effective photoisomerization (response) time:

$$\tau = \frac{\tau_c}{1 + \tau_c(\sigma_t q_t I + \sigma_c q_c I)}$$

Thus, for large I, τ is inversely proportional to intensity:

$$\tau = \frac{1}{(\sigma_t q_t + \sigma_c q_c)I}$$

An order-of-magnitude estimate of τ is 0.04 sec (agreeing well with Fig. 5), assuming $q_t \sim 0.7$, $q_c \sim 0.3$, $\sigma_t \sim 10000$ cm^{-1}, $\sigma_c \sim 1000$ cm^{-1}, $I \sim 8.9 \times 10^{18}$ photons/(cm^2 sec), and density $= N_A/(250$ g/mol), where N_A is Avogadro's number.

Plots of the detector signal V corresponding to different input power levels are shown in Figure 6 in the case $r = 1$, for a planar cell. Note that the phase transition takes place only if $r \geq 1$.

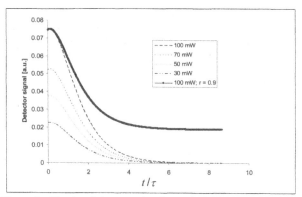

Figure 6: Simulated detector signal as a function of t/τ, for comparison with Figure 5. Simulated thickness was 1.4 μm.

CONCLUSIONS

    Adding properly dispersed conductive nanoparticles (MWCNTs and SWCNTs) to azo LCs modifies not only the absorption of the cell, but also the dynamics of a cw laser transmitting through the sample, dramatically decreasing the response time of twist cells. We have shown that conductive carbon nanotubes may enable all-optical switching technology, using LCs. We have developed a theory to quantitatively characterize the nonlinear transmission process of a system of crossed polarizers comprising the LC cell.

ACKNOWLEDGMENT

The authors acknowledge Nano-C Inc. of Westwood, MA (USA) for providing proprietary carbon nanotube inks.

REFERENCES

1. I. C. Khoo, M. V. Wood, M. Y. Shih, P. H. Chen, Optics Express **4**, 431, 1999.
2. N.V. Tabiryan, S.V. Serak, V.A. Grozhik, J. Opt. Soc. Am. B, **20** (3), 1-7, 2003.
3. S. V. Serak, N.V. Tabiryan, T. J. Bunning, **16** (4), 471-483, 2007.
4. U. Hrozhyk, *et. al.*, Proc. of SPIE **7050**, 705007 (1-11), 2008.
5. G. Cook, *et. al.*, Opt. Express, **16**, 4015-4022, 2008.
6. C. Khoo, Kan Chen and Y. Zhang Williams, IEEE JSTQE **12** (3), 443, 2006.
7. N.V. Kamanina, Opto-Electronics Review, **12**(3), 285-289, 2004.
8. C. Khoo, et. al., Appl. Phys. Letts. **82**, 3587, 2003.
9. I. C. Khoo, Optics Letters, **20**, 2137 (1996).
10. H. Duran, B. Gazdecki, A. Yamashita, T. Kyu, Liquid Crystals, **32**(7), 815, 2005.
11. Y. Wang, K. Kempa, B. Kimball, J. Carlson, G. Benham, W. Z. Li, T. Kempa, J. Rybczynski, A. Herczynski, Z. F. Ren, Appl. Phys. Letts., **85**, 8607, 2004.

12. U. A. Hrozhyk, S. V. Serak, N. V. Tabiryan, L. Hoke, D. M. Steeves, B. Kimball, G. Kedziora, Mol. Cryst. Liq. Cryst., **489**, 257, 2008.
13. U. Hrozhyk, S. Serak, N. Tabiryan, T.J. Bunning, Mol. Cryst. Liquid Cryst., **454**, 235, 2006.
14. Svetlana V. Serak, Nelson V. Tabiryan, Proc. of SPIE **6332**, 63320Y (1-13), 2006
15. P.G. de Gennes, Physics of Liquid Crystals, Clarendon Press, Oxford, 1974.

Mater. Res. Soc. Symp. Proc. Vol. 1142 © 2009 Materials Research Society   1142-JJ10-06

Sorting Carbon Nanotubes by Their Diameter and Number of Walls

Sungwoo Yang[1] and Jie Liu[1]

[1]Chemistry Department, Duke University, 124 Science Drive, Durham, NC 27708, U.S.A.

ABSTRACT

Double walled carbon nanotubes (DWNTs) were synthesized by chemical vapor deposition method using a metal mixture of cobalt and molybdenum, supported by titanium silicalite (TS-1). Ethanol was a carbon feeding source. After purifying steps, including oxidation in the air and treatment with NaOH and the major species of as-synthesized CNTs were DWNTs. Well dispersed CNT solutions were achieved using the surfactant sodium cholate. The density gradient separation method using ultracentrifugation was used to sort CNTs by their diameter and the number of sidewalls. UV-vis NIR, Raman, Photoluminescence (PL) and Transmittance electron microscopy (TEM) were studied to distinguish various diameter and different types of CNTs. These studies were used to demonstrate sorting of synthesized CNTs by their diameter and number of sidewalls.

INTRODUCTION

Carbon nanotubes (CNTs) have fascinated numerous scientists over the last two decades due to their outstanding material properties, such as high electrical conductivity and Young's modulus. These electronic and mechanical properties vary with the type of CNT depending on the diameter and the number of sidewalls. Over the last decade, many methods have been developed to control the type of CNT and its diameter to some degree. However, structural heterogeneity in terms of diameter and the number of sidewalls is still unavoidable. Therefore, the separation of CNTs in terms of diameter and the number of sidewalls is one of the great challenges to realize CNT's applications, including CNT composites and transparent conductive films. Recently, several methods of sorting CNTs by their diameter have been reported [1-6]. However, sorting CNTs by their number of sidewalls has not yet been reported. Hereby, we report sorting CNTs by their diameter and number of sidewalls, such as single and double walled carbon nanotubes (SWNTs and DWNTs). In our report, PL, UV-vis NIR, Raman and TEM were used to study different types of CNT.

EXPERIMENT

(1) Impregnation catalyst with TS-1 zeolite
The Impregnation method was used to produce catalysts. Co and Mo were used as metal seeds in the same manner as the combustion method. However, zeolites were used as a support, because their specific porosity encourages DWNT production. Co, Mo and zeolites with a specific molar ratio (Co:Mo = 6:1) were dissolved into methanol and heated on the hot plate to

remove all of the solvents. The resulting solid phase mixture was annealed at 530°C to remove residual organic impurities. Lastly, final impregnation catalysts were obtained for CVD growth of CNTs.

(2) Well-suspended CNT solution with surfactants
First, carbon nanotubes were synthesized using TS1 catalysts in a specific molar ratio of Co:Mo and reaction temperature. (Be Specific)(900°C, 1 hour) After these raw CNTs were purified by air oxidation and HF treatment, they were added into deionized (DI) water with surfactants (2 % w/v, Sodium Cholate). This combination was thoroughly mixed by tip-sonication (at 45% of 750 Watts, 20 KHz for 45 minuates).

(3) Chemical density gradient
A chemical density gradient was created by mixing iodixanol in DI water (from 20 % w/v to 50 % w/v). The higher concentration of iodixanol solution went to the bottom of the separation tube. Well-suspended CNTs were then added on top of this density gradient solution.

(4) Density gradient ultra-centrifugation
The tube containing density gradient solution with suspended CNTs was centrifuged at 150,000G for twelve hours to separate CNTs by their diameter and number of side walls. Each layer was carefully decanted and analyzed by several methods, such as PL, UV-vis, Raman spectroscopy and TEM to determine their electric and mechanic properties.

(5) Transparent CNT thin film
Sorted CNT solutions by density gradient ultra-centrifugation were filtered using isopre membrane. After filtration, CNT thin films were transfered to a glass, and any residual isopre membrane was removed by washing with chloroform. Film thickness was measure by Zygo optical system, and four point resistivity system measured CNT film's conductivity.

DISCUSSION

(1) TEM anaysis
After density gradient centrifugation, surfactants were removed from separated CNTs to take TEM images by washing in ethanol. As shown in figure 1, CNTs in the upper layer region are mostly small diameter SWNTs. On the other hand, relatively large DWNTs were the major species of the bottom layer region. Differences in the buoyant densities of SWNTs and DWNTs result in this separation.

TEM

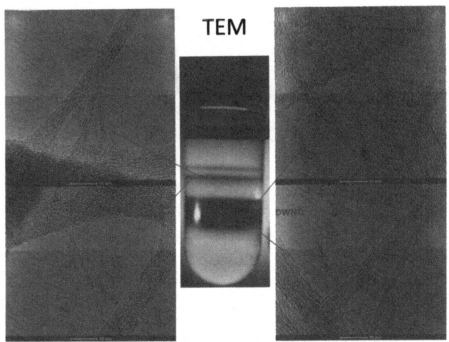

Figure 1: TEM images. Left one is corresponding to the top layer, and the right one is bottom layer.

(2) Raman analysis

Radial vibrations in carbon nanotubes give rise to the radial breathing mode (RBM) stretches in the Raman spectra of CNT; it is well known that the diameter of CNT is inversely proportional to its Raman RBM shift. In this manner, the diameter of each separated sample was estimated. According to Kitaura's plot, three different wavelength lasers (442 nm, 633 nm and 785 nm) can be used to detect CNT's with diameters ranging between 0.5 nm and 2.7 nm. As shown in figure 2, lower layer include larger diameter of CNT. In addition, the lower layer shows both inner and outer walls of the CNT, and their diameter difference was around 0.75 nm.

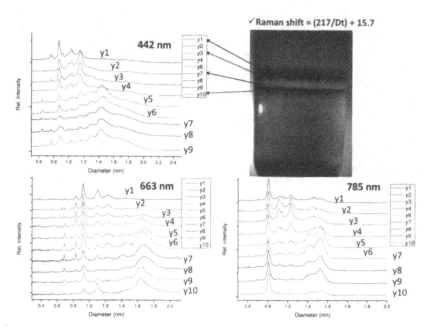

Figure 2: Raman RBM. Raman shifts were converted to diameter of CNTs. y1 corresponds to the 1ˢᵗ layer and y2 does to the 2ⁿᵈ layer and so fourth. Bottom layer includes large diameter of CNT.

The graphite band (G-mode) directly indicates the presence of CNT, and shows two main features (G- and G+ components) observed with frequencies at ~1570 and ~ 1590 cm-1. These two features are associated with transverse (TO phonon) and longitudinal (LO phonon) vibration along the nanotube. As shown in figure 3, Raman G- mode has decreased in the lower layer, and with 785 nm wavelength layer, the Raman G+ mode was clearly shifted to the lower frequency.

118

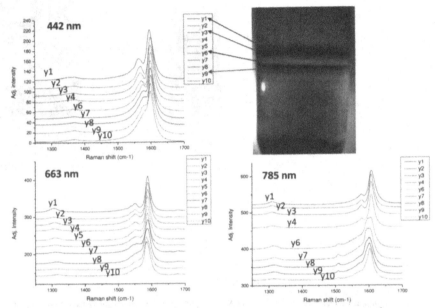

Figure 3: Raman G-mode. Theree different wavelenght layers were used. y1 corresponds to the 1ˢᵗ layer and y2 does to the 2ⁿᵈ layer and so fourth. At the bottom layer, Raman G⁻ mode has decreased.

(3) Photoluminescence (PL) analysis

Optical transitions of carbon nanotube are rather sharp and strong. Consequently, it is possible to selectively excite nanotubes which have certain (m, n) chiral vector. This enables us to estimate (n,m) chiral vectors of nanotubes using PL. In this experiment, separated CNTs were excited at 650 nm wavelength laser. As a result, (8,3), (7,5) and (7,6) nanotubes were observed with specific emission wavelength. As shown in figure 4, the 6th layer includes larger diameter nanotubes compared to the 1st layer. In addition, PL quenching was observed in layers which correspond to DWNTs. This observation might be a evidence of inner-tube interaction of DWNT as non-radiative photon relaxation.

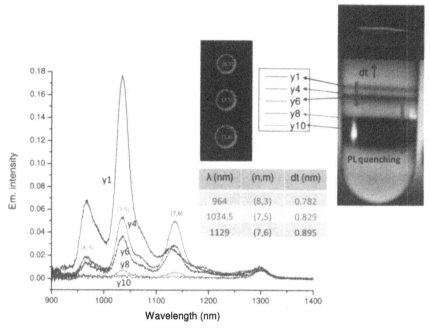

Figure 4: Photoluminescence (PL) of separated CNTs. The 6th layer (y6) include bigger SWNTs compared to the 1st layer. In addition, PL quenching was observed in DWNT layers.

(4) Absorption spectroscopy (UV-vis NIR)

The absorption provides information for the resonant peaks originating from van Hove singularities. Therefore, this UV-vis NIR spectroscopy enables us to evaluate (n,m) chiral vector of nanotubes. As shown in figure 5, the 6th layer contains bigger diameter CNT compared to the 1st layer, and both the 1st and the 6th layers correspond to SWNT regions. On the other hand the 9th, 12th and 14th layers show absorption spectroscopy of DWNT. These DWNT absorption peaks in the 1st transitional energy (~ 900 nm and ~ 1250 nm) become wider compared to SWNT's absorption peaks.

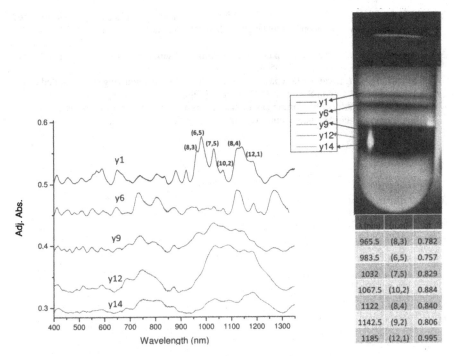

Figure 5: Photoluminescence (PL) of separated CNTs. The 6th layer (y6) include bigger SWNTs compared to the 1st layer. In addition, PL quenching was observed in DWNT layers.

CONCLUSIONS

This study demonstrates that carbon nanotubes can be sorted by their diameter and the number of walls using density gradient centrifugation. As a result of this separation, it was possible to study the Raman, PL, and UV-vis NIR exclusively of DWNTs . Separated DWNT show the diminishment of the Raman G- mode and PL quenching. In addition, DWNT absorption peaks were achieved.

REFERENCES

1. Zheng, M. *et al.* Structure-based carbon nanotube sorting by sequence-dependent DNA assembly. Science 2003, 302, 1545–1548
2. Chen, Z. H. *et al.* Bulk separative enrichment in metallic or semiconducting single-walled carbon nanotubes. Nano Lett. 2003, 3, 1245–1249.

3. Krupke, R., Hennrich, F., Kappes, M. M. & Lohneysen, H. V. Surface conductance induced dielectrophoresis of semiconducting single-walled carbon nanotubes. Nano Lett. 2004, 4, 1395–1399.

4. Maeda, Y. *et al*. Large-scale separation of metallic and semiconducting single-walled carbon nanotubes. J. Am. Chem. Soc. 2005, 127, 10287–10290.

5. Maeda, Y. *et al*. Large-scale separation of metallic and semiconducting single-walled carbon nanotubes. J. Am. Chem. Soc. 2005, 127, 10287–10290.

6. Michael S. Arnold. *et al*, Sorting carbon nanotubes by electronic structure using density differentiation. Nature Nanotechnology. 2006, 1, 60 – 65.

Mater. Res. Soc. Symp. Proc. Vol. 1142 © 2009 Materials Research Society   1142-JJ10-21

Theoretical Study of Single Graphene-like Semiconductor Layer Made of Si and Transition Metal Atoms

Takehide Miyazaki[1], Noriyuki Uchida[2] and Kanayama Toshihiko[2]
[1]Research Institute for Computational Sciences, National Institute for Advanced Industrial Science and Technology,
AIST Tsukuba Central 2, Umezono 1-1-1, Tsukuba 305-8563, Japan
[2]Nanodevice Innovation Research Center, National Institute for Advanced Industrial Science and Technology,
AIST Tsukuba West 7, Onogawa 16-1, Tsukuba 305-8569, Japan

ABSTRACT

Based on first-principles atomic structure optimization, we demonstrate that a single layer of Si atoms in graphene-like positions may become semiconducting upon attachment of transition metal atoms with six valence electrons such as Mo, W and Cr. The resultant chemical formula is $(MSi_6)_n$ where M is either of Mo, W or Cr. We predict the indirect energy gaps, which are ~0.3 eV for M=Mo and ~0.2 eV for M=Cr and W with the generalized gradient approximation to the density functional theory. We find corrugation of the Si layer in the film-normal direction to occur for all of the three transition metal elements investigated. A possible cause of energy gap opening is discussed.

INTRODUCTION

A recent trend in downsizing of field effect transistors (FETs) is pointing sub-10-nm regimes, where the physical gate length and channel thickness should be ~6 nm and less than ~2nm, respectively[1]. The fabrication of those aggressively miniaturized FETs should be performed with atomic scale accuracies. This has prompted researchers to envisage bottom-up approaches that introduce new materials with one-dimensional and two-dimensional anisotropies.

Graphene is a typical example of two-dimensional materials, which has been extensively studied as a candidate for the channel layer in the ultra-scaled FETs [2]. It is semimetallic in nature while the FET channel regions should be semiconducting. Although several experiments have demonstrated that graphene bilayers can be semiconducting by the application of electric field [3,4] or by cutting into nanoribbons [5,6], it is worth attempting to search for other layered materials, particularly those mainly composed of Si.

A goal of this theoretical study is to propose a conceptual possibility of Si-based ultra-thin semiconductor layers used for nano-scale electronic devices. It should be kept in mind that we do not intend to find the global energy minimum of Si-based nanostructures of any shape, but find ultra-thin film structures with semiconducting nature. Based on this spirit, we have so far studied Si thin films resembling the double graphene layers, in which transition metal atoms are intercalated. We have shown that $(MoSi_{12})_n$ and $(ZrSi_{12})_n$ have the energy gaps of ~0.5 eV and ~0.3 eV, respectively, in generalized gradient approximation (GGA) to density functional theory (DFT) [7,8]. The structure of these films are characterized by the MSi_{12} building blocks (M:

transition metal atom) which have a hexagonal prism shape with the M atom inside of the prism cavity [9].

In the present study, we show that it is possible to make a semiconducting state even in a single graphene-like Si layer, provided that the transition metal atoms with six valence electrons such as Mo, Cr and W are attached to the Si layer. For example, we have found that a $(MoSi_6)_n$ film has the indirect GGA band gap of ~0.3 eV. The structure is obtained by removing a six-member ring of Si in the unit cell of $(MoSi_{12})_n$ mentioned above and relaxing the positions of all remaining atoms. The thickness of this film is ~0.2 nm, if the atomic radii of Si and Mo are not taken into account. Three highest valence and three lowest conduction bands across the energy gap are both characterized by the mixture of the d components of Mo and the p components of Si_6. This suggests that the gap opening is mainly due to a substantial overlap among these orbitals.

METHOD OF CALCULATION

We optimize the stable geometries of the $(MSi_6)_n$ crystals, by using the STATE computer code that performs first-principles electronic structures[10] on the basis of the density-functional theory (DFT)[11,12] within the generalized gradient approximation (GGA) [13]. The film structures are modeled in supercell geometries for which periodic boundary conditions are applied to not only in the directions parallel to the film [x and y directions in Fig.1(a)] but also in the film-normal direction [z direction in Fig.1(a)]. In the x and y directions, a regular triangular Bravais lattice is assumed. The periodicity in the z direction is set to be 1.27nm.

The ultra-soft[14,15] (for Mo, Cr, W) and norm-conserving[16] (for Si) pseudopotentials are used to describe only the valence electronic structures. The plane-wave basis set is used to expand the Bloch functions, with the energy cutoff being 25 Ry. The augmentation functions used for the ultrasoft pseudopotentials are pseudized according to the recipe in Ref.[15] and expanded in plane waves up to 400 Ry. Self-consistent field (SCF) iterations are continued until the total energy converges to ~3×10^{-5} eV/atom. The positions of all Si atoms in a supercell are relaxed within the C_{3v} symmetry constraint around each M atom until the magnitude of the residual force acting on each atom becomes less than ~0.26 eV/nm. The integration over the irreducible wedge of the first Brillouin zone (1BZ) of a supercell is performed with k mesh equivalent to the 12X12X1 k points in the whole 1BZ.

RESULTS

Atomic structures

We have determined the structures of $(MSi_6)_n$ in the following way. (i) The initial configurations were prepared by removing one Si layer from the double-Si-layer $(MoSi_{12})_n$ reported in our previous study[8]. (ii) The Mo atom was replaced by the M atom (this step was skipped in case of M=Mo), where the generated topology is like that shown in Fig.1. (iii) The lattice constant and the internal structural degrees of freedom were simultaneously optimized within the C_{3v} symmetry constraint around the M atom by setting a_1 and a_2 equal to each other and also γ=60 degrees (see Fig.1 for definitions of these parameters). Next, we have checked whether the film structures are stable against small distortions to break the C_{3v}

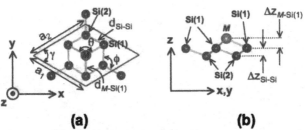

(a)   **(b)**

Figure 1. Structure models of $(MSi_6)_n$ film, where M is either of Mo, W and Cr, viewed from the film normal [(a)] and parallel [(b)] directions. In panel (a), a rhombus is a unit cell. The structural parameters for each M are listed in **Table I**.

symmetry. For this purpose, we repeated an extensive set of structure optimizations in $(MoSi_6)_n$ by varying the lattice constant a_1 [a_2 was fixed at the value a_0=0.655 nm obtained in the step (iii)] and the angle γ ranging from -4% to +4% around a_0 and 60 degrees, respectively. The evaluated dataset was fitted in the 4th-order polynomials of a_1 and γ. We found that the interpolated total energy surface is convex and has a single minimum at a_1=a_0 and γ=60 degrees. Thus we conclude that the $(MoSi_6)_n$ film structure with the C_{3v} symmetry is at least a *local* minimum. It is highly expected that the structures of $(WSi_6)_n$ and $(CrSi_6)_n$ are also the cases because of their similar structures of the Si layers and the M valency to those of $(MoSi_6)_n$.

We illustrate the atomic structure of $(MSi_6)_n$ (M=Mo, W, and Cr) in Fig. 1. The optimized structural parameters are listed in Table I. The overall structural features are the following: (i) The Si layer is corrugated in the direction normal to the film (the z direction, see Fig. 1 for definition of the axes) in the sense that the two sublattices in the Si layer, one with Si(1) atoms and the other with Si(2) atoms are displaced in the z direction with the separation Δz_{Si-Si} [Fig.1(b)]. (ii) The M atom is positioned at the hollow site of the Si six-member ring.

In more detail, the structure of the Si layer is contracted in the x and y directions and expanded in the z direction relative to the (111) bilayer in bulk Si. The magnitude of the corrugation (Δz_{Si-Si}) is 0.083 nm ~ 0.087 nm, larger than the thickness of the (111) bilayer in bulk Si, 0.0784 nm. Correspondingly, the bond angles among the Si(1)-Si(2)-Si(1) atoms (ϕ, see also Fig.1 for definition of the atom notations) range from 104.6 degrees (M=W) to 16.6 degrees (M=Mo), which are all less than the counterpart in the diamond structure, 109.5 degrees. The corrugation is not favored by the Si layer alone but caused by the M attachment to the hollow site. In summary, the above structural features (i) and (ii) both contribute to the energy gain since the bonding interactions between the M d orbital and the Si sp hybrids and among the latter themselves are simultaneously enhanced.

Table I. Optimized structure parameters of $(MoSi_6)_n$, $(WSi_6)_n$, and $(CrSi_6)_n$ for equal a_1 and a_2(a_1= a_2=a_0) and γ= 60 degrees (see Fig.1 for the notations of the structure parameters).

compound	a_0 (nm)	d_{Si-Si} (nm)	$d_{M-Si(1)}$ (nm)	θ (degs)	ϕ (degs)	$\Delta z_{M-Si(1)}$ (nm)	Δz_{Si-Si} (nm)
$(MoSi_6)_n$	0.655	0.232	0.245	103.4	106.6	0.104	0.083
$(WSi_6)_n$	0.659	0.233	0.244	104.6	104.6	0.097	0.087
$(CrSi_6)_n$	0.656	0.230	0.235	106.2	106.2	0.088	0.085

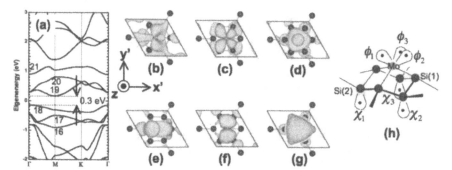

Figure 2. (a) Electronic energy band structure of $(MoSi_6)_n$. The valence-band maximum (VBM) is located in the 18-th band near Γ point. The conduction-band minimum (CBM) occurs in the 19-th band at M point. Panels (b)-(g): isosurfaces (0.001 per unit cell) of squared amplitudes of the eigenstates for 16th and 17th bands at M [(b) and (c)], 18th band at Γ [(d)], 19th and 20th bands at M [(e) and (f)], and 21st band at Γ [(g)], respectively. The x' and y' axes are rotated from the x and y axes (Fig.1) by 30 degrees. (h) Schematic illustration of the dangling bonds of Mo (ϕ_1, ϕ_2 and ϕ_3) and Si(2) (χ_1, χ_2 and χ_3) atoms. A lobe with a dot represents a dangling bond with an unpaired electron.

Electronic structure of $(MoSi_6)_n$

Hereafter, we mainly focus our attention on the electronic structure of $(MoSi_6)_n$. The energy bands of $(MoSi_6)_n$ along the high symmetry axes in the first Brillouin zone are shown in Fig.2(a). The indirect GGA band gap of ~0.3 eV opens between the 18th and 19th energy bands. The band gaps of $(WSi_6)_n$ and $(CrSi_6)_n$ are both ~0.2 eV. The valence band maximum is located near (not exactly at) Γ point of the 18 th band. The conduction band minimum occurs at M point of the 19th band.

We take a closer look at the six bands across the energy gap, the three (bands #16, #17 and #18) occupied and the other three (bands #19, #20 and #21) unoccupied. Main contributions of the Mo d states to these bands may be clarified if their eigenstates at high symmetry k-points Γ and M are analyzed in the x' and y' Cartesian coordinate axes (see Fig.2). In order to identify the character of each state, we analyzed isosurfaces not only shown in Fig.2 but also those plotted from various directions combined with analysis of the atom-projected density of states (PDOS, not shown in this paper). The Mo components in the 16th and 17th bands at M point are mainly of the $d_{x'^2-y'^2}$ (perturbed with some $d_{x'z}$ component) [Fig.2 (b)] and $d_{x'y'}$ characters [Fig.2 (c)], respectively. The 18th band, the highest valence band, has a large $d_{3z^2-r^2}$ component of Mo at Γ point [Fig.2 (d)]. The 19th (lowest conduction band) and 20th bands at M point mainly exhibit the $d_{x'z}$ (perturbed with some $d_{x'^2-y'^2}$ component) [Fig.2 (e)] and $d_{y'z}$ characters at Mo [Fig.2 (f)], respectively. The 21st band at Γ point shows the mixed s and p_z characters of Mo [Fig.2 (g)]. In the PDOS result, the Si p_z character is clearly observed in the occupied (#16, #17 and #18) bands and in the unoccupied 21st band while the Si p_x and p_y components are much weaker in all of the six bands (#16 ~ #21). Further, the Si p_z character in the three occupied bands (#16, #17 and #18) is stronger at Si(2) than Si(1).

126

DISCUSSION

Here we discuss a qualitative origin of energy gap opening in $(MoSi_6)_n$. In a simple electron counting picture, a cause of the energy gap can be regarded as termination of six dangling bonds (DBs) of the Si_6 unit with six valence electrons of a Mo atom. First of all, the DBs at three Si(1) atoms are terminated with three Mo valence electrons via formation of the Mo-Si(1) bonds. The corresponding electronic states are included in the energy bands #10 through #15 according to the PDOS. Therefore, the bands #16 ~ #21 shown in Fig.2 (a) mainly come from the DBs at three Si(2) atoms and the remaining three Mo valence electrons. In order to understand how these electronic states interact, let us begin with a situation where each of the DBs at the Mo atom (ϕ_1, ϕ_2 and ϕ_3) and Si(2) atoms (χ_1, χ_2 and χ_3) accommodates an unpaired electron [Fig.2 (h)]. The symmetries of the Mo orbitals shown in Figs.2 (e), (f) and (g) may be derived from linear combinations of the Mo DBs, $2\phi_1 - \phi_2 - \phi_3$, $\phi_2 - \phi_3$ and $\phi_1 + \phi_2 + \phi_3$, respectively[17]. This suggests that the bands #19, #20 and #21 originate from Mo DBs. Since these bands are actually unoccupied, one should understand that the electrons at the Mo DBs are transferred to Si(2) atoms to fulfill the DBs χ_1, χ_2 and χ_3. In fact, one can see the substantial contribution of the p_z character at Si(2) in Figs.2 (b), (c) and (d), respectively. The Si(2) p_z orbitals are aligned with the Mo sp_z hybrid in the anti-bonding manner [Fig.2 (g)]. This may explain why the 21st band is the highest in energy among the six bands #16 ~ #21. In summary, combination of the Mo-Si(1) bond formation and charge transfer from Mo to Si(2) is a cause of the energy-gap opening in $(MoSi_6)_n$. Similar arguments apply to $(WSi_6)_n$ and $(CrSi_6)_n$.

In light of our material design toward nano-scale transistors, the most valuable message from the present theoretical results is a *conceptual support* to the use of transition metal atoms as Si DB killers in ultra-thin Si films. In fact, the Si films synthesized via coalescence of the $MoSi_x$ (x~10) clusters do exhibit semiconductor properties [18].

Since our current interest is in a possible mechanism with which the Si DBs are terminated with transition metal atoms, we do *not* insist that a freestanding, and completely flat slab form as shown in Fig.1 is the global energy minimum among all possible atomic configurations of the $(MSi_6)_n$ compound. There might be some possibility of the film scrolling in tubular forms or bowl-like shapes, because the structure asymmetry in the z direction might induce the z-dependence of the stresses in the x and y directions. We are now investigating whether this actually happens. In literature similar cases have already been reported. Prinz et al.[19] and Schmidt and Eberl[20] have demonstrated the synthesis of nanotubes by rolling up the nanosheets, composed of a one monolayer (1 ML) GaAs that is attached to a 1ML InAs ultra-thin film and SiGe, respectively, utilizing the stress relaxation caused by the asymmetric film structure. The asymmetry in the film structure may also be induced by adsorption as theoretically proposed by Yu and Liu[21], who have found that scrolling of a graphene sheet occurs if hydrogen atoms could be adsorbed on one side of the graphene surface.

CONCLUSION

We propose that a single layer of Si atoms in graphene-like topology may be semiconducting if the transition metal atoms of six valence such as Mo, W and Cr are attached to one side of the Si layer. The GGA band gaps of ~0.3 eV, ~0.2 eV and ~0.2 eV are predicted for M=Mo, W and Cr, respectively. The Si layer is corrugated in the z direction, becoming similar to a bilayer of Si(111). The energy gap opens because the six Si dangling bonds in the unit cell are efficiently terminated with six valence electrons of M by making substantial overlap with the M d orbitals and also by transfer of the electrons of M to the Si dangling bonds.

ACKNOWLEDGMENTS

This research was partially supported by Scientific Research on Priority Areas of New Materials Science Using Regulated Nano Spaces, the Ministry of Education, Science, Sports and Culture, Grant-in-Aid for 472, 19051017, 2007. A part of computation has been performed on the "T2K-Tsukuba" supercomputer system. The atomic structures and isosurfaces are drawn using the VESTA[22].

REFERENCES

1. International Technology Roadmap for Semiconductors, SIA, EECA, EIAJ, KSIA, TSIA, 2007.
2. K. S. Novoselov, A. K. Geim, S. V. Morozov, D. Jiang, Y. Zhang, S. V. Dubonos, I. V. Grigorieva and A. A. Firsov, *Science* **306**, 666 (2004).
3. T. Ohta, A. Bostwick, T. Seyller, K. Horn, and E. Rotenberg, *Science* **313**, 951 (2006).
4. J. B. Oostinga, H. B. Heersche, X. Liu, A. F. Morpurgo, and L. M. K. Vandersypen, *Nature Mat.* **7**, 151 (2008).
5. X. Wang, Y. Ouyang, X. Li, H. Wang, J. Guo, and H. Dai, *Phys. Rev. Lett.* **100**, 206803 (2008).
6. X. Li, X. Wang, L. Zhang, S. Lee, and H. Dai, *Science* **319**, 1229 (2008).
7. T. Miyazaki and T. Kanayama, *Mater. Res. Soc. Proc.* **0958**, L01-07 (2007).
8. T. Miyazaki and T. Kanayama, *Appl. Phys. Lett.* **91**, 082107 (2007).
9. H. Hiura, T. Miyazaki and T. Kanayama, *Phys. Rev. Lett.* **86**, 1733 (2001).
10. "Simulation Tool for Atom Technology (STATE)"; See, Y. Morikawa, H. Ishii and K. Seki, *Phys. Rev. B* **69**, 041403(R) (2004).
11. P. Hohenberg and W. Kohn, *Phys. Rev.* **136**, B864 (1964).
12. W. Kohn and L. J. Sham, *Phys. Rev.* **140**, A1133 (1964).
13. J. P. Perdew, K. Burke and M. Ernzerhof, *Phys. Rev. Lett.* **77**, 3865 (1996).
14. D. Vanderbilt, *Phys. Rev. B* **41**, 7892 (1990).
15. K. Laasonen, A. Pasquarello, R. Car, C. Lee and D. Vanderbilt, *Phys. Rev. B* **47**, 10142 (1993).
16. N. Troullier and J. L. Martins, *Phys. Rev. B* **43**, 1993 (1991).
17. T. A. Albright, J. K. Berdett and M.-H. Whangbo, *"Orbital Interactions in Chemistry"*, (John Wiley & Sons, 1985), pp.381-385.
18. N. Uchida, H. Kintou, Y. Matsushita, T. Tada, and T. Kanayama, *Appl. Phys. Exp.* **1**, 121502 (2008).
19. V. Ya. Prinz, V. A. Seleznev, A. K. Gutakovsky, A. V. Chehovskiy, V. V. Preobrazhenskii, M. A. Putyato, and T. A. Gavrilova, *Physica E* **6**, 828 (2000).
20. O. G. Schmidt and K. Eberl, *Nature* **410** 168 (2001).
21. D. Yu and F. Liu, *Nano Lett.* **7** (2007) 3046.
22. K. Momma and F. Izumi, *J. Appl. Crystallogr.* **41**, 653 (2008).

Mater. Res. Soc. Symp. Proc. Vol. 1142 © 2009 Materials Research Society   1142-JJ10-28

Single crystalline boron nanocones

X. J. Wang[1,a], J. F. Tian[1], L. H. Bao[1], C. M. Shen[1], H. J. Gao[1]

[1]Beijing National Laboratory for Condensed Matter Physics, Institute of Physics,
Chinese Academy of Sciences, Beijing 100080, People's Republic of China
[a] Present address: Department of Electronic Engineering, University of Electro-Communications, 1-5-1
Chofugaoka, Chofu-Shi, Tokyo,182-8585, Japan

Abstract: Single crystalline boron nanocones have been prepared by a simple chemical vapor deposition method using thermal evaporation of B/B_2O_3 powders precursors in an Ar/H_2 gas mixture at the synthesis temperature of 1000-1200 °C and Fe_3O_4 nanoparticles as catalyst. The length of boron nanocone is about 5 micrometers, and the diameter of the nanocone tip is about 50 nm. TEM and HRTEM indicate that the nanocone is single crystalline boron. The results of field emission measurement show the low turn-on and threshold electric fields of about 3.5 V/μm and 5.3 V/μm at the vacuum gap of 200 μm. Boron nanocones with good field emission properties are promising candidates for applications in optical emitting devices and flat panel displays.

Keywords: Boron nanocone, Single crystal, Field emission

1. Introduction

Boron is a unique element with structure complexity due to the electron-deficient nature of boron. Extensive research of bulk boron has been reported. Recent experimental results show that boron may become metallic under a pressure of about 160 GPa at room temperature, and transform from a nonmetal to a superconductor at 6 K. The critical temperature of transition increases from 6 K to 11.5 K further increasing the pressure up to 250 GPa.[1] In the past decade, the theoretical and experimental studies about one-dimensional boron nanomaterials have been also extensively developed. Some theoretical studies have suggested that the existence of layered, tubular and fullerene-like boron solids, which possess many novel structural, electronic and thermal properties. The proposed boron nanotubes possess a metallic-like density of states, which are predicted to be relative stable and have metallic conductivities exceeding those of carbon nanotubes with potential applications in field emission and high-temperature electron devices.[2] These new boron nanotubes, if synthesized, may have potential applications in nanoelectronics, in field emission device and in nanocomposites where they may impart stiffness, toughness, and strength.

Despite the importance of boron nanotubes, there were little reports on the experimental synthesis of boron nanotubes, except a few reports on the synthesis of boron nanowires and nanobelts. Amorphous boron nanowires were first fabricated by chemical vapor transport method using boron, iodine and silicon as precursor, and by radio-frequency magnetron sputtering of the target made of boron and boron oxide.[3,4] Crystalline boron nanowires were synthesized by chemical vapor deposition reaction of diborane (B_2H_6) in Ar gas on the alumina substrates.[5] In recent years, crystalline boron nanobelts have been also synthesized using laser ablation of high-purity boron pellets and by pyrolysis of B_2H_6 at low temperature of 630-750 °C.[6] However, as another important one-dimensional boron nanomaterial, few report about boron nanocones was obtained although it is expected that one-dimensional nanocones should have some better properties than that of nanowires and nanobelts.

In this paper, we prepared the boron nanocones by a simple chemical vapor deposition method, the thermal evaporation of B/B_2O_3 powders precursors in an Ar/H_2 gas mixture at the reaction temperature of 1000-1200 °C. Raman spectrum and field emission properties of boron nanocones were measured to explore the future possibility applications in optical emitting devices.

2. Results and discussion

Figures 1(a) show the TEM image of boron nanocone grown on Si(111) substrate at 1100 °C for 2 h. The length of nanocone is 5 micrometers. The diameter of the nanocone reduces gradually from about 300 nm at the root to the range of 50 nm at the tip. Figure 1(b) shows a HRTEM image of nanocone. The boron nanocone is crystalline and covered by 1-2 nm thick amorphous layer. The preferential growth direction was determined to be the [001] direction. The outer amorphous layer is an oxide sheath, and formed when the crystalline boron nanocones were exposed in air.[7,8]

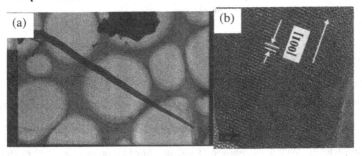

Figures 1(a) and (b) TEM image and HRTEM images of boron nanocone

Raman spectrum properties of boron nanocones

Figure 2 shows the Raman spectrum of boron nanocones film. It can be found that approximately 10 broad bands with halfwidths up to 40 nm were observed from Raman spectrum. This is similar to Raman spectrum of pure bulk boron, however, which is much less than 83 photons allowed in first order Raman scattering expected by group theory. It is suggested that many phonon peaks are buried in these broad structures and are overlapping by a considerably high damping.[9]

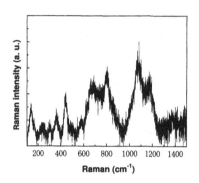

Figure 2. Raman spectrum of boron nanocones film

Field emission properties of boron nanocones film

Figure 3(a) shows the current density versus the electric field (J-V) characteristics for the boron nanocones at the different vacuum gaps of 200 μm, 300 μm and 500 μm. Electric fields at the current densities of 10 μA/cm^2 and 1 mA/cm^2 are defined to be the turn-on and the threshold electric fields, respectively. With a vacuum gap of 200 μm, low turn-on and threshold electric fields of about 3.5 V/μm and 5.3 V/μm were observed for the boron nanocones. The emission current density was increased to 10mA/cm^2, and no saturation tendency was seen with the electric field increased to 9.5 V/μm. When the vacuum gap was increased from 200 μm to 500 μm, the threshold electric field decreased monotonously from 5.3 V/μm to 3.3 V/μm, while the voltage applied on the two electrodes increased from about 1060 V to 1650 V. The threshold field value is slightly higher than that reported for aligned carbon nantubes (1.6 V/μm),[10] close to that of ZnO nanopins (5.9 V/μm),[11] but lower than that of aligned ln$_2$O$_3$ nanowires (10.7 V/μm).[12]

Figure 3(b) shows the corresponding F-N plots of boron nanocones at the different vacuum gaps from 200 μm to 500 μm. It exhibits a typical characteristic of field emission from semiconductors. Similar nonlinear characteristics of F-N plots were also found in In$_2$O$_3$ semiconductor nanowires grown on InAs substrate.[12] The field emission stability of the boron nanocones film was performed at a vacuum gap of 200 μm. An electric field of 9.5 V/μm was applied at a chamber pressure of 8×10^{-7} Pa for 6 h. The boron nanocones exhibit a stable long-term emission without obvious fluctuation at relative low field emission current density, as shown in Fig. 3(c). The less than ±5% current fluctuation meets the field emission criterion for display.

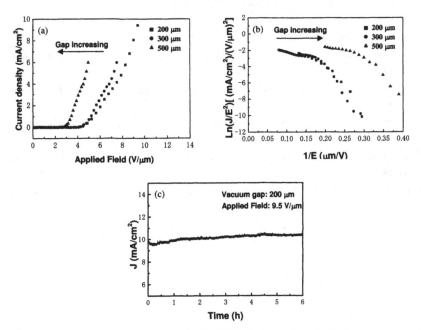

Figure 3. (a) The current density versus the electric field (J-V) characteristics for the boron nanocones at the vacuum gaps of 200 μm, 300 μm and 500 μm. (b) The corresponding F-N plots. (c) The field emission stability of the boron nanocones at the vacuum gap of 200 μm and applied field of 9.5 V/μm for 6 h.

3.Conclusions

Single crystalline boron nanocones have been fabricated by a simple chemical vapor deposition method. The results of field emission measurement show low turn-on and threshold electric fields of about 3.5 V/µm and 5.3 V/µm at the vacuum gap of 200 µm. When the vacuum gap was increased to 500 µm, the threshold electric fields were decreased to 3.3 V/µm.

References

[1]Eremets M, Struzhkin V, Mao H K and Hemley R 2001 *Science* **293** 272

[2]Boustani I and Quandt A 1997 *Europhys. Lett.* **39** 527

[3]Wu Y Y, Messer B and Yang P D 2001 *Adv. Mater.* **13** 1487

[4]Cao L M, Zhang Z, Sun L L, He M, Wang Y Q, Li Y C, Zhang X Y, Li G, Zhang J and Wang W K 2001 *Adv. Mater.* **13** 1701

[5]Otten C J, Loune O R, Yu M F, Cowley J M, Dyer M J, Ruoff R S and Buhro W E 2002 *J. Am. Chem. Soc.* **124** 4564

[6]Wang Z K, Shimizu Y, Sasaki T, Kawaguchi K, Kimura K and N. Koshizaki 2003 *Chem. Phys. Lett.* **368** 663

[7] Wang X J, Tian J F, Yang T Z, Bao L H, Hui C, Liu F, Shen C M, Xu N S, and Gao H J 2007 *Adv. Mater.* **19** 4481

[8] Wang X J, Tian J F, Bao L H, Hui C, Yang T Z, Liu F, Shen C M, Xu N S, and Gao H J 2008 *Chin. Phys.* **17** 1647

[9]Richter W, Hausen A, Binnenbruck H 1973 *Phys. Stat. Sol B*, **60** 461

[10]Rao A M, Jacques D, Haddon R C, Zhu W, Bower C, Jin S 2000 *Appl. Phys. Lett.* **76** 3813

[11]Xu C X, Sun X W 2003 *Appl. Phys. Lett.* **83** 3806

[12]Li S Q, Liang Y X, Wang T H 2006 *Appl. Phys. Lett.* **88** 053107-1

Mater. Res. Soc. Symp. Proc. Vol. 1142 © 2009 Materials Research Society   1142-JJ10-33

Tailoring Conductivity of Carbon Nanotubes Network by Halogen Oxoanions

Seon-Mi Yoon,[1] Sung Jin Kim,[2] Hyeon-Jin Shin,[1,2] Anass Benayad,[1] Jong Min Kim,[1] Young Hee Lee[*,2] and Jae-Young Choi[*,1]

[1]Samsung Advanced Institute of Technology, Samsung Electronics CO., LTD, Suwon 440-600, Korea
[2]Department of Physics, Department of Nanoscience and Nanotechnology, and Center for Nanotubes and Nanostructured Composites, Sungkyunkwan Advanced Institute of Nanotechnology, Sungkyunkwan University, Suwon 440-746, Korea

ABSTRACT

Chlorine oxoanions with the chlorine atom at different oxidation states were introduced in an attempt to systematically tailor the electronic structures of single-walled carbon nanotubes (SWCNTs). The degree of selective oxidation was controlled systematically by the different oxidation state of the chlorine oxoanion. Selective suppression of the metallic SWCNTs with a minimal effect on the semiconducting SWCNTs was observed at a high oxidation state. The adsorption behavior and charge transfer at a low oxidation state were in contrast to that observed at a high oxidation state. These results concurred with the experimental observations from X-ray photoelectron spectroscopy. The conversion of metallic SWCNTs to semiconducting ones at high oxidation state might result in an increase in the resistance of the intrinsic SWCNTs but reduces the contact resistance significantly by decreasing the number of metal-semiconductor junctions. The sheet resistance of the SWCNT film decreased significantly at high oxidation states, which was explained in terms of a p-doping phenomenon that is controlled by the oxidation state.

INTRODUCTION

The modification of the atomic and electronic structures of carbon nanotubes is a crucial step in many applications.[1] Strong chemical bonding between the host materials and carbon nanotubes is essential for enhancing the mechanical properties of the host materials, whereas a simple dispersion of nanotubes in a host matrix can enhance the conductivity of the host materials.[2] The functionalization of single-walled carbon nanotubes (SWCNTs) by hydrogenation and fluorination leads to a significant change not only in the atomic structures but also in the electronic structures of carbon nanotubes.[3,4] Doping control of carbon nanotubes by various chemical means is also strongly dependent on the types of chemical dopants.[5] Various dispersants have been used to disperse carbon nanotubes. This often involves serious modification of the electronic structures of the carbon nanotubes.[1] The presence of functional groups in the dispersant may induce a permanent or an induced dipole moment in a molecular solvent. This presumably involves charge transfer between the absorbates and carbon nanotubes, which modifies the electronic structures of the carbon nanotubes. In order to tailor the electronic structures of carbon nanotubes to a desired direction, it is essential to understand the effect of the absorbates.
Halogen oxoanions are a systematic oxidant in engineering redox reactions due to the existence of different oxidation states. As the oxidation state of the halogen oxoanions increases,

the number of electrons participating in its redox reaction increases, whereas the redox potential of the reaction decreases.[6] Most oxidants can be used as a p-dopant in carbon nanotubes. For example, $SOCl_2$, iodine, and a simple acid treatment enhance the majority carrier concentration and increase the conductivity of carbon nanotubes (CNTs).[7-10] This trend can be altered depending on the physisorption or chemisorption states of the adsorbates. Nevertheless, an understanding of the oxidation of CNTs and their related physical and chemical properties are still unclear. Chlorine oxoacids can provide useful information on the doping of CNTs in this matter, which can be engineered by the different oxidation states with the same chlorine atoms. In general, chlorine atoms induce weaker charge transfer than other halogen atoms.[6] Therefore, the effect of oxidation by chlorine oxoanions can be correlated exclusively to the electronic structures of the CNTs.

The aim of this study was to determine how the electronic structures of CNTs can be engineered systematically by oxidants. In this article, chlorine oxoanions with different oxidation states were introduced to correlate the degree of oxidation with the electronic structures of the CNTs. Raman spectroscopy showed that metallic SWCNTs were selectively doped at a high oxidation state without any significant modification of the semiconducting SWCNTs. In addition, the degree of doping was controlled by the oxidation state. X-ray photoelectron spectroscopy (XPS) demonstrated that physisorption was involved at a high oxidation state, whereas chemisorption played an important role in modifying the electronic structures of the CNTs at a low oxidation state

EXPERIMENT

The SWCNTs were synthesized by arc discharge and purchased from Iljin Nanotech Co. Ltd.. The chloro oxoanion sodium salts ($NaClO_x$) were dissolved in deionized water to produce 20 % aqueous solution. To compare effect of oxidation state, this aqueous $NaClO_x$ solution of 0.0161 M was further dissolved in 10 ml of 1-Methyl-2-pyrrolidinone (NMP) solution. One mg of the SWCNTs was then added to the mixed solution and sonicated in a bath type sonicator (BANDELIN SONOREX) at 240 W for 10 hours. This solution was filtered through an anodisc filter (Whatman Ltd.) with a pore size of 0.1 μm. The CNTs film was dried at room temperature overnight. The sheet resistance at room temperature was measured using a four-point probe (AIT Co. Ltd., SR1000N). The CNT films were characterized by micro-Raman spectroscopy (Renishaw RM1000-Invia). Two excitation energies of 2.41 eV (514 nm, Ar+ ion laser), 1.96 eV (632.8 nm, He-Ne laser) with a Rayleigh line rejection filter, which accepts a spectral range of $50 \sim 3200$ cm^{-1} were used in this study. XPS analysis (QUANTUM 2000, Physical electronics) using a focused monochromatized Al Ka radiation (1486.6 eV) was carried out to check for the presence of residual material and the degree of doping effect.

DISCUSSION

Chlorine oxoanions are commonly available halogen species with various oxidation states. At a low oxidation state, the redox potential is high, thus the reaction is thermodynamically favored. The number of electrons involves in the redox reaction, together with redox potential can be correlated exclusively to the doped state of the CNTs.

Figure 1 shows the Raman spectra of various samples treated with chlorine oxoanions. At an excitation energy of 633 nm, both metallic and semiconducting SWCNTs are excited in the pristine sample. The G-band is deconvoluted into one BWF line shape, one metallic and four semiconducting components in Lorentzian shape.[11] The presence of BWF component (left-shaded region) of G-band that represents a long energy tail at lower energy side in the pristine sample is an evidence for the existence of metallic SWCNTs in the sample.[12] This metallic component decreased gradually with increasing oxidation state, and the BWF line profile disappeared at oxidation states of +5 and +7. The two main peaks in RBMs represented the abundance of the metallic nanotubes. These peaks were gradually reduced with increasing oxidation states, in consistent with BWF changes. On the other hand, at an excitation energy of 514 nm, the semiconducting SWCNTs were excited exclusively, i.e., neither the BWF line shape in G-band nor the metallic component in the RBMs is observed. In this case, there were no appreciable changes in G-band and RBMs, which was in contrast with those observed at 633 nm.

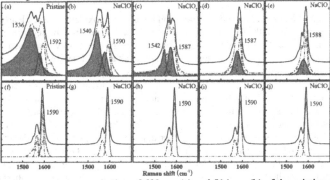

Figure 1. Raman spectra at an excitation of 633 nm (a) and 514 nm (b) of the pristine and chloro oxoanion treated samples (16.1mM) at different oxidation states before heat treatment.

This successive modification of the metallic component with increasing oxidation state can be understood by the fact that electrons are extracted from the metallic CNTs to the adsorbates through oxidation. Nevertheless, the peak position of the G^+-band near 1592 cm^{-1} at 633 nm, which represents the semiconducting nanotubes, was downshifted at an oxidation state of +1 and +3. In this case, the electrons were transferred from the adsorbates to the semiconducting CNTs. This is in contrast to the typical oxidation effect, which showed the uptake of electrons and the corresponding upshift in the G^+-band.[8] These observation reflects the general rule that a semiconductor has larger electron affinity than a metal.[13] This might be related to the fact that the high redox potential at low oxidation state can promote a strong interaction with the CNTs, which acts as a rate-limiting step for further oxidation. On the other hand, no peak shift in the G^+-band was observed in the case of samples with exclusively semiconducting SWCNTs at 514 nm. Therefore, the metallic SWCNTs reacted selectively with the chlorine oxoanions, particularly at a low oxidation state, and were converted to semiconducting ones with pseudo bandgap opening,[14] whereas the semiconducting SWCNTs were quite inert to these oxidants over the entire range of oxidation states.

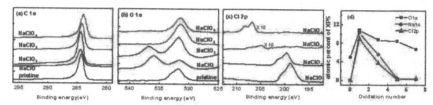

Figure 2. XPS analysis of (a) C1s, (b) O1s, and (c) Cl2p core peaks and (d) the atomic percentage of O1s (square), Na1s (circle), and Cl2p (triangle) determined from quantitative analyses.

XPS analysis was performed to understand the effect of oxidation more clearly (Figure 2). A thorough analysis of the C1s core peak provides valuable information regarding the nature of carbon bonding. For the pristine CNT film, the C1s spectrum is mainly composed of a large sp^2 carbon component at 284.3 eV, a small sp^3 carbon peak at 285.1 eV, oxygen related groups, and π-π* plasmon satellite. [15] The most component of C1s spectra in the all samples treated by oxidants was about sp^2 carbon. This means that the oxidation doesn't involve defects on CNTs. In the case of oxidation by NaClO and NaClO2, a new spectrum appeared at approximately 534~538 eV in O 1s spectrum, which was assigned to chemisorbed oxygen and/or water. However, the precise assignment of the O1s peak at higher binding energies presents some difficulties due to the presence of a sodium Auger line in this energy range.[16] No such peaks were observed at the high oxidation states. The Cl2p peak observed near 199 eV was attributed to the presence of salt in the samples (NaCl). This was confirmed by the Na1s peak around ~1072 eV (not shown here). In case of NaClO2, another peak near 1073 eV was observed, which is due to the presence of sodium oxide.[16] Thus, the new peak in O 1s should indicate the presence of a sodium to compensate charge on surface of CNTs. This result was confirmed by quantitative analysis which the amount of residual element was increasing at lower oxidation state.(Figure 2d)

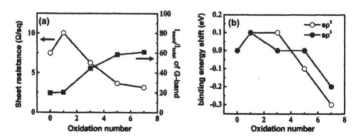

Figure 3. (a) Sheet resistance (open circle) and the relative areal intensity (close square) of semiconducting components with respect to the whole areal intensity of the G-band at 1.96 eV and (b) binding energy shift of sp^2 and sp^3 of C1s XPS in terms of the oxidation number.

In order to determine how this selective modification of metallic SWCNTs affects the sheet resistance, the sheet resistance was measured as a function of the oxidation state (Figure

3a). The sheet resistance was increased by approximately 30 % at an oxidation state of one, and decreased further with increasing oxidation state. The relative areal intensity of the semiconducting components was also extracted by deconvoluting the G-band (Figure 3a). The relative areal intensity was increased with increasing oxidation state. In general, the sheet resistance of the random network SWCNTs is determined by the sum of resistances of the intrinsic SWCNT network and tube-tube contact. The tube-tube contact is composed of metal-metal and semiconductor-semiconductor junctions that give ohmic behavior, and a metal-semiconductor junction that forms a Schottky barrier.[17] The conversion of metallic SWCNTs to semiconducting ones at high oxidation state might result in an increase in the resistance of the intrinsic SWCNTs but reduces the contact resistance significantly by decreasing the number of metal-semiconductor junctions. Nevertheless, this argument is not applicable at an oxidation state of one, i.e., the resistance increased compared with the pristine sample. Figure 3b shows the binding energy shift of C1s measured by XPS. Both peaks related to sp^2 and sp^3 carbon were upshifted by approximately 0.1 eV at the low oxidation states of +1 and +3, whereas they were downshifted at the high oxidation states, which is the case with a typical downshift in p-doped CNTs.[7,10] The atomic percentage of sodium and chlorine increased significantly at an oxidation state of +1(Figure 2d). This may be the origin of the upshift of C1s, i.e., the charges are transferred to CNTs from adsorbates. Thus the p-type carrier is compensated by an effective n-doping, resulting in the increase of sheet resistance as observed in Figure 3a. On the other hand, there was no appreciable sodium and chlorine remaining at the high oxidation states of +5 and +7. This suggests that the reaction process involves a complicated pathway of adsorption and decomposition into the final products such as sodium chloride and oxygen species. Generally, increasing number of oxygen in oxoanion, the oxoanion interacts weakly on CNT due to the steric hindrance and the reaction involved adsorption and reduction. Thus, this results in the different reaction pathway of oxoanions on CNTs depending on oxidation state. At low oxidation state, the strongly adsorbed chemicals remain on CNT after redox reaction in contrast with at high oxidation state.

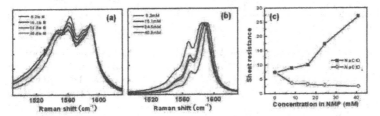

Figure 4. Raman spectra at an excitation of 633 nm of NaClO (a) and NaClO$_4$ (b) treated samples and sheet resistance(Ω/sq)(c) of the samples as function of concentration

This charge transfer was also dependent on the concentration of the oxidants as shown in Figure 4. In case of NaClO, the metallic component decreased slightly and G$^+$ peak position was not shifted with increasing concentration. However in case of NaClO$_4$, the metallic component disappeared at 8.2mM and the G$^+$ peak position was downshifted in comparison with the pristine as shown in Figure 1. The G$^+$ peak position also was upshifted gradually with increasing concentration. This suggests that the reaction is influenced by metallicity of SWCNTs and concentration of oxidant. At low concentration, the metallic SWCNTs release charges to the adsorbates, whereas the semiconducting SWCNTs extract charges from the adsorbates, as evidenced from the Raman spectra. On the other hand, at high concentrations, both metallic and semiconducting SWCNTs release charges to the adsorbates. At higher concentration of $NaClO_4$, semiconducting SWCNTs were also p-doped and the sheet resistance decreases like semiconductor. But in case of $NaClO$, the sheet resistance was increased depending on the concentration. This should be influenced by the residual element on CNT. An optimum concentration is needed to selectively modify the metallic SWCNTs with a minimum change in the semiconducting SWCNTs. These modification could change the electronic properties.

CONCLUSIONS

This study examined the effect of oxidation on CNTs by introducing chloro oxoanions with the chlorine atom at various oxidation states. At a low concentration of oxidants with a high oxidation state, selective oxidation was observed on the metallic carbon nanotubes with a minimal effect on the semiconducting nanotubes, where the charges are extracted from mainly from the metallic nanotubes to the adsorbates. At a low concentration of oxidants, the oxidation behavior at a low oxidation state was quite different from that observed at a high oxidation state. The oxidation at a low oxidation state involves stronger adsorption and charge transfer from the adsorbates to CNTs. On the other hand, oxidation at a high oxidation state involves an intermediate physisorption that evolves to dissociate into further byproducts of the oxygen adsorbates, leaving charge transfer from the CNTs to adsorbates. The sheet resistance was reduced significantly when the p-doping behavior was dominant. The finding of selective doping on metallic SWCNTs with oxidation can be applied to the precise doping control of several electronic devices.

REFERENCES

1. D. Tasis, N. Tagmatarchis, A. Bianco, M. Prato, *Chem. Rev.* **106**, 1105 (2006).
2. Y. Hu, O. Shenderova, Z. Hu, C.W. Padgett, D. W. Brenner, *Rep. Prog. Phys.* 1847 (2006).
3. A. Nikitin, H. Ogasawara,D. Mann, R. Denecke, Z. Zhang, H. Dai, K. Cho, A. Nilsson, *Phys. Rev. Lett.* **95**, 225507 (2005).
4. H. Peng, P. Reverdy, V. N. Khabashesku, J. L. Margrave, Chem. Comm. 362 (2003).
5. (a) J. Chen, M. A. Hamon, H. Hu, Y. Chen, A. M. Rao, P. C. Eklund, R. C. Haddon, Science **282**, 95 (1998). (b) C. Zhou, J. Kong, E. Yenilmez, H. Dai, *Science* **290**, 1552 (2000).
6. D. F. Shriver, P. W. Atkins, *Inorganic Chemistry* (Oxford University Press, 1999) Chapter 12, pp 405-429.
7. U. Dettlaff-Weglikowska, V. Skákalová, R. Graupner, S. H. Jhang, B. H. Kim, H. J. Lee, L. Ley, Y. W. Park, S. Berber, D. Tománek, S. Roth, *J. Am. Chem. Soc.* **127**, 5125 (2005).
8. A. M. Rao, P. C. Eklund, S. Bandow, T. Thess, R. E. Smalley, *Nature* **388**, 257 (1997).
9. M. E. Itkis, S. Niyogi, M. E. Meng, M. A. Hamon, H. Hu, R. C. Haddon, *Nano Lett.* **2**, 155 (2002).
10. H.-Z. Geng, K. K. Kim, K. P. So, Y. S. Lee, Y. Chang, Y. H. Lee, *J. Am. Chem. Soc.* **129**, 7758 (2007).
11. S. D. M. Brown, A. Jorio, P. Corio, M. S. Dresselhaus, G. Dresselhaus, R. Saito, K. Kneipp, *Phys. Rev. B.* **63**, 155414 (2001).

12. A. Jorio, A. G. Souza Filho, G. Dresselhaus, M. S. Dresselhaus, A. K. Swan, M. S. Unlu, B. B. Goldberg, M. A. Pimenta, J. H. Hafner, C. M. Lieber, R. Saito, *Phys. Rev. B* **65**, 155412 (2002).
13. D. Chattopadhyay, I. Galeska, F. Papadimitrakopoulos, *J. Am. Chem. Soc.* **125**, 3307 (2003).
14. M. S. Strano, C. A. Dyke, M. L. Usrey, P. W. Barone, M. J. Allen, H. Shan, C. Kittrell, R. H. Hauge, J. M. Tour, and R. E. Smalley, *Science* **301**, 1519 (2003).
15. (a) F. R. McFeely, S. P. Kowalczyk, L. Ley, R. G. Cavell, R. A. Pollak, D. A. Shiriley, *Phys. Rev. B* **9**, 5268 (1974). (b) J. Diaz, G. Paolicelli, S. Ferrer, F. Comin, *Phys. Rev. B* **54**, 8064 (1996).
16. J. F. Moulder, W. F. Stickle, P. E. Sobol, K. D. Bomben, *Handbook of X-ray Photoelectron Spectroscopy* (Physical Electronics, Inc. 1992).
17. M. S. Fuhrer, J. Nygard, L. Shih, M. Forero, Y. G. Yoon, M. S. C. Mazzoni, H. J. Choi, J. S. Ihm, S. G. Louie, A. Zettl, P. L. McEuen, *Science*, **288**, 494 (2000).

Mater. Res. Soc. Symp. Proc. Vol. 1142 © 2009 Materials Research Society 1142-JJ10-37

Geometrical Factor Dependent Brewster Angle Shift of Silicon Nano-Pillars

Fan-Shuen Meng, Yi-Hao Pai, and Gong-Ru Lin*

Institute of Photonics and Optoelectronics, and Department of Electrical Engineering,

National Taiwan University

No. 1 Roosevelt Road Sec. 4, Taipei 106, Taiwan R.O.C.

*grlin@ntu.edu.tw

ABSTRACT

The aspect ratio dependent optical reflectivity and Brewster angle shift of Si nano-pillars with different geometrical sizes, in which the height is increasing from 65 nm to 210 nm and the diameter is enlarging from 50 nm to 90 nm is investigated. These Si nano-pillars were fabricated by introducing thermally self-aggregated Ni nano-dots on SiO_2 buffered Si substrates, and used them as a mask to dry-etch the Si substrate. The Si nano-pillars greatly reduce the reflectivity from 0.34 to 0.066 as the Si nano-pillars height increases from 65 to 210 nm in visible light region. The Brewster angle shifts from 73.4° to 62.2° as the Si nano-pillars height increases from 65 to 105 nm. When the height and pedestal of Si nano-pillars become greater than 135 nm and 50 nm, the Brewster angle phenomenon observed for the TM mode reflectivity completely diminishes.

INTRODUCTION

Since 1980, Lowdermilk has demonstrated that graded-index antireflection (AR) surfaces may widely reduce insertion losses at the interfaces between different optical media [1]. Graded-index AR surfaces has been widely studied because that AR coatings used for the mass production on different surfaces are associated with problems such as adhesion, thermal mismatch, and the stability of the thin-film stack [2,3]. The graded-index AR surfaces could greatly reduce the surface refractivity for a wide spectral bandwidth and over a large field of view by introducing a continuous effective index gradient between the substrate and the surrounding medium [4]. For fabricating the graded-index AR surfaces, researchers were usually preparing a periodical subwavelength-antireflective-structured (SAS) array. The periodicity of this SAS array is smaller than the wavelength of the incident light. But if we prepared the graded-index AR surfaces by fabricating SAS array, we should use so many expensive processes just like e-beam lithography that the product will not be manufactured for commercialization [3,5]. In 2007, G. R. Lin et al. fabricated Si nano-pillars array by using a Ni nano-dot mask assistant reactive-ion-etching process. This method was just only utilizing metal evaporation, RTA process and ICP-RIE process to prepare graded-index AR surfaces [6]. By using this method, the products based on graded-index AR surfaces will be probably for commercialization. In this work, by changing the parameters of Ni nano-dot mask assistant reactive-ion-etching process, different geometrical sizes of Si nano-pillars array can be obtained, which contribute to different optical characteristic in reflectivity and Brewster angle shift.

EXPERIMENTAL DETAILS

The Si nano-pillars array was fabricated on n-type (100)-oriented Si substrate by employing Ni/SiO_2 nano-mask assistant dry-etching procedure, as schematically shown in Fig. 1. In this procedure, a 50-nm-thick Ni film was evaporated on the SiO_2 /Si substrate using an e-beam evaporating system with Ni deposition rate of 0.1 Å/ s under an applied current of 70 mA. The SiO_2 buffered layer with a thickness of 200 Å is deposited by using plasma enhanced chemical vapor deposition under standard recipe. Subsequently, a rapid thermal annealing process at 850 °C for 22 s under the N_2 flowing gas of 5 SCCM (SCCM denoted cubic centimeter per minute of STP) is performed to self-aggregate the Ni nano-dots on SiO_2 /Si substrate.

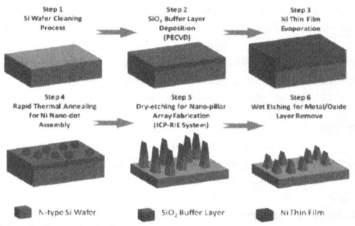

Fig.1 Schematic diagram for fabricating Si nano-pillars array by using Ni/SiO_2 nano-mask assistant dry-etching procedure on Si substrate is illustrated.

By using the Ni nano-dots with area density of 5×10^{11} cm^{-2} as a nano-mask, the Si substrate is dry etched in a planar-type inductively coupled plasma–reactive ion etching (ICP-RIE) system (SAMCO ICP-RIE 101iPH) at RF frequency of 13.56 MHz and the ICP/bias power conditions of 100/50 Watts. The etching gases of CF4 and Ar mixture with a condition of $CF_4/Ar=40/40$ SCCM were introduced into the reactive chamber through individual electronic mass flow controllers. The chamber pressure of 0.66 Pa remains unchanged during an etching duration of 5–7 min [6]. The Si nano-pillars array can be prepared by detuning the etching time to obtain different height and aspect ratio. Fig. 2 clearly shows that we can fabricate a series of samples with different height of nano-pillars from 65 to 210 nm. The density of Si nano-pillars can be detuned by the density of Ni nano-dots, however, we remain the Si nano-pillar density as constant during our sample preparation in order to clarify the effect of Si nano-pillar height on the surface reflectance. The wavelength dependent reflectivity spectra were measured by multiple wavelength reflectivity measurement system. The angular dependent reflectivity spectra were measured by self-founded spectroscopy system.

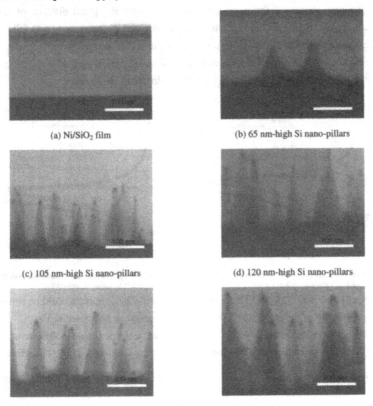

(a) Ni/SiO₂ film

(b) 65 nm-high Si nano-pillars

(c) 105 nm-high Si nano-pillars

(d) 120 nm-high Si nano-pillars

(e) 135 nm-high Si nano-pillars

(f) 210 nm-high Si nano-pillars

Fig.2 SEM photographs of Si nano-pillars array with different size of height (a) Ni/SiO₂ film, (b) 65 nm high, (c) 105 nm high, (d) 120 nm high, (e) 135 nm high, (f) 210 nm high which obtained by detuning the etching duration.

DISCUSSION

From wavelength dependent reflectivity spectroscopy which was shown in the Fig. 3 and Fig. 4, the reflectivity spectra present ripple in UV region for both TM and TE modes due to the shallow optical penetration depth induced Fabry-Perot interference effect. In visible light region, the Si nano-pillars greatly reduce the reflectivity from 0.34 to 0.066 as the Si nano-pillars height increases from 65 to 210 nm, which is verified by TM and TE mode reflectivity at small incident angle. The effective refractive index of Si nano-pillars is found to reduce from 3.147 to 1.499 by increasing the nano-pillars height from 65 to 210 nm. We also experimentally found that the Si nano-pillars exhibit lower reflectivity than the bulk Si and SiO_2 thin film by introducing an infinite reflective interface with gradient refractive index. Since each reflective light comes from a different depth from the interface, each will have a different phase. If the transition takes place over an optical distance of $\lambda/2$, all phases are present, there will be destructive interference and the reflectivity will fall to zero [4]. This statement told us that the reflective light comes from different interface can always find another light comes from another layer with $\lambda/2$ optical path difference to induce destructive interference just like the concept used in single slit interference.

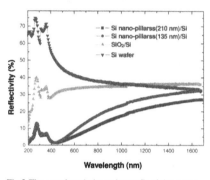 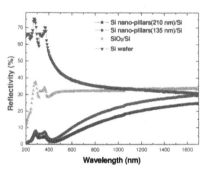

Fig.3 The wavelength dependent reflectivity spectra was measured at TE mode polarization.

Fig.4 The wavelength dependent reflectivity spectra was measured at TM mode polarization.

The intersection of TM- and TE-mode dependent reflectance spectra could be attributed to the discontinuity of the structure for Si nano-pillars array, as illustrated in Fig. 5. When TM-mode light was incident, the component of electric field upon the nano-pillars is larger than the condition of incident by TE mode. Although such difference did not significantly affects in shorter (< 460 nm) and longer (> 860 nm) wavelength regions, which has made the

TE- and TM-mode reflectivities of Si nano-pillars array reversely (TM > TE mode) in between. It is possibly that the energy of shorter wavelength light is sufficiently large to strongly drive electrons for TE mode regardless of the discontinuity of Si nano-pillars array. When illuminating long-wavelength light, the nano-pillar surface behaves more like a uniform one, which is due to a significant difference between the spacing of nano-pillars and the wavelength of the incident light. With the wavelength ranging between 460 and 860 nm, the energy of incident light was not enough to strongly drive electrons on nano-pillars. The component of electric field upon the nano-pillars began to dominate the change of reflectivity. The nano-scaling spacing of nano-pillars still performed as a discontinue material since the spacing is comparable with the wavelength in this region. Such a phenomenon is quite different from common case.

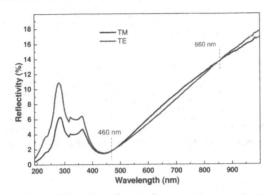

Fig.5 The cross phenomenon of wavelength dependent reflectivity for TM and TE mode.

The angular dependent spectrum is measured by homemade reflectrometry system with two coaxial rotation stages and substitutable laser sources at versatile wavelengths. The photodetector is mounted upon the outer rotation stage and the sample holder is fixed on the axis of the inner stage. The angular dependent spectrum is taken by concurrently rotating the inner and outer stages for θ degree and 2θ degree, respectively. From the angular dependent diagnosis of TE and TM polarized optical reflectivity at 780 nm, as illustrated in Fig. 6. The Brewster angle of TM mode reflectivity for bulk Si is at 75°, which is significantly reduced by surface roughening the Si wafer with nano-pillars. The Brewster angle shifts from 73.4° to 62.2° as the Si nano-pillar height increases from 65 to 105 nm. With increasing nano-pillar height, the contrast ratio of reflectivity at Brewster angle to that at lower angles is gradually decayed from 47.02 to 1.47. When the height of Si nano-pillars become greater than 135 nm, the Brewster angle phenomenon observed for the TM mode reflectivity completely diminishes while leaving a greatly reduced reflectivity at small incident angles for Si nano-pillar roughened surface. In addition, both the TE and TM mode reflectivity of 135 nm–high Si nano-pillar decrease by 50% as compared to that of the silicon wafer which further decay to 25% of its original reflectivity if the nano-pillar height increases to 210 nm.

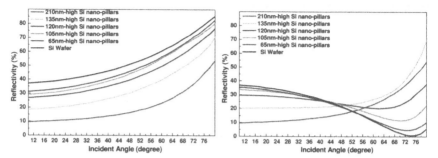

Fig.6 The angular dependent reflectivity spectra of TE (right) and TM (left) mode.

CONCLUSION

In conclusion, detuning the etching duration and RF power of nano-mask assistant dry-etching process to prepare different samples with different geometric factors about height of Si nano-pillars and spacing is reported. This randomly SAS arrays exhibit many optical properties different from uniform material. The reflectivity spectra present ripple in UV region for both TM and TE modes. In visible light region, the reflectivity greatly was reduced from 0.34 to 0.066 as the Si nano-pillars height increases from 65 to 210 nm. The effective refractive index is reduced from 3.147 to 1.499 by increasing the nano-pillars height from 65 to 210 nm. The wavelength dependent reflectivity spectra exhibits a cross phenomenon of TE and TM polarized light between 460 to 860 nm. The Brewster angle of TM mode reflectivity shifts from 73.4° to 62.2° by surface roughening with nano-pillars from 65 to 105 nm. With increasing nano-pillar height, the contrast ratio of reflectivity at Brewster angle to that at lower angles is gradually decayed from 47.02 to 1.47. The Brewster angle phenomenon of TM mode reflectivity completely diminishes as the height exceeding 135 nm. These characteristics may be applied in many different optical devices in future.

ACKNOWLEDGMENTS

The authors thank the National Science Council of Republic of China and the Excellent Research Projects of National Taiwan University for financially supporting this research under grants NSC 97-ET-7-002-007-ET, NSC 97-2221-E-002-055 and 97R0062-1.

REFERENCES

1. H.W. Lowderiviilk, and D. Millaai, Laser Focus, pp. 64 (1980).
2. S. Walheim et al., Science, vol. 283, 520 (1999).
3. Y. Kanamori, M. Sasaki, and K. Hane, Opt. Lett., vol. 24, pp. 1422 (1999).
4. S. J. Wilson and M. C. Hutley, Opt. Acta., vol. 29, pp. 993 (1982).

5. Y. Zhaoning, G. He, W. Wei, G. Haixiong, and Y. C. Stephen, Am. Vac. Sci., pp. 2874 (2003).

6. Lin, Gong-Ru, Chang, Ya-Chung Chang, Liu, En-Shao, Kuo, Hao-Chung, and Lin, Huang-Shen, Appl. Phys. Lett., vol. 90, pp. 181923 (2007).

Mater. Res. Soc. Symp. Proc. Vol. 1142 © 2009 Materials Research Society   1142-JJ10-38

Self-aligned growth of single-walled carbon nanotubes using optical near-field effects

Y. S. Zhou, W. Xiong, M. Mahjouri-Samani, W. Q. Yang, K. J. Yi, X. N. He and Y. F. Lu*
Department of Electrical Engineering, University of Nebraska-Lincoln, Lincoln, NE 68588-0511

ABSTRACT

By applying optical near-field effects in a laser-assisted chemical vapor deposition (LCVD) process, self-aligned growth of ultra-short single-walled carbon nanotubes (SWNTs) was realized in a well controlled manner at a relatively low substrate temperature due to the nanoscale heating enhancement induced by the optical near-field effects. Bridge structures containing single suspending SWNT channels were successfully fabricated. Ultra-sharp tip-shaped metallic electrodes were used as optical antennas in localizing and enhancing the optical fields. Numerical simulations using High Frequency Structure Simulator (HFSS) reveal significant enhancement of electrical fields at the metallic electrode tips under laser irradiation, which induces localized heating at the tips. Numerical simulations were carried out to optimize SWNT growth conditions, such as electrode tip sharpness and film thickness, for maximal enhancement of electrical near field and localized heating.

INTRODUCTION

Single-walled carbon nanotubes (SWNTs) are ideal platforms for fundamental study and promising candidates for technological applications due to their unique one-dimensional structures and superior properties [1]. SWNTs have been intensively investigated as an alternative material to replace current silicon-based devices [2]. Fabrication of SWNT-based devices requires integration of SWNTs. Chemical vapor deposition (CVD) process is presently the only economically viable process for integrating SWNTs into devices by yielding selective and aligned SWNT growth directly on various substrates [3]. Conventional CVD methods generally require high reaction temperatures up to 800 to 1200 °C, and expose the substrate to the high temperature. However, high substrate temperature prevents integration of SWNTs into device fabrication.

It is well known that ultra-sharp metallic tips can be used as optical antennas in localizing and enhancing optical fields due to the optical near-field effects [4]. Enhancement of the electric field at the tip results in an enhanced eddy current, which yields local heating enhancement and a significant temperature increase at the tip. In this study, optical near-field effects are integrated in the LCVD process to realize the self-aligned growth of SWNTs at a relatively low substrate temperature. The experimental results and numerical simulations demonstrate that the

---

* Correspondence should be addressed to Prof. Y. F. Lu, Tel: (402) 472-8323, Fax: (402) 472-4732, Email: ylu2@unl.edu.

laser-induced optical near-field effects produce nanoscale localized heating at electrode tips, which stimulates the selective growth of SWNTs at the tips. Influence of the tip sharpness and metallic film thickness were investigated via numerical simulation using HFSS to optimize conditions for maximal enhancement of the localized electric field and temperature increase. Two obvious advantages are observed for this developed technique: precise position control and low substrate temperature for SWNT growth.

EXPERIMENT

Figure 1. Schematic LCVD experimental setup for SWNT growth using optical near-field effects.

SWNTs were directly grown on pre-patterned SiO_2/Si substrates via the LCVD process. Figure 1 shows the schematic LCVD experimental setup for SWNT growth using the optical near-field effects. Enhanced optical fields are strictly localized at electrode tips, as indicated in Fig. 1. Heavily doped p-type Si wafers covered with a 2-μm-thick oxide layer were used as the substrates. Electrode patterns were fabricated by conventional lithography followed by DC sputtering. Metallic films consisting of 200 nm Ruthenium (Ru) and 2 nm iron (Fe) were deposited. Ru was chosen as the electrode material due to its low solubility of Fe, high work function, and high melting point. The Fe thin films were used as the catalyst for SWNT growth. Ultra-sharp tip-shaped electrodes were fabricated by focused ion beam (FIB) nano-machining using an FEI Strata 200xp system. The electrode gap distance was controlled to be around 200 nm. Growth of SWNTs was carried out in an LCVD chamber. A continuous-wave (CW) CO_2 laser (Synrad, firestar v40, wavelength 10.6 μm) was used to irradiate the substrates. A DC voltage about 1.0 V/μm was applied between the electrodes to assist the self-aligned growth of the SWNTs. A gas mixture of acetylene (C_2H_2) and anhydrous ammonia (NH_3) with a volume ratio of 1:10 was introduced into the chamber. Anhydrous ammonia is used as a buffer gas in the LCVD process to dilute acetylene, create an etching environment to suppress the growth of amorphous carbon, and protect catalyst particles from being poisoned by amorphous carbon. The gas flow was directed to blow from the cathode to the anode to assist the self-aligned growth of SWNTs. The reaction pressure was maintained at 10 Torr. The reaction temperature was controlled around $550\,^{\circ}C$, which was $150\,^{\circ}C$ lower than that required in conventional CVD processes. A voltmeter was connected into the circuit to monitor the growth process of SWNTs. The reaction process was terminated immediately when a SWNT-bridge was formed.

Scanning electron micrographs were taken using a Hitachi S4700 field-emission scanning electron microscope (SEM) with an ultimate resolution of 1.2 nm at 25 kV. Electrical transport measurements were taken on an Agilent HP4145B semiconductor parameter analyzer with an ultimate measurement resolution of 100 pA. Raman spectra were taken on a home-made micro-Raman spectrometer with an Argon laser (Coherent, Innova 300, 514.5 nm) as the excitation source. The spatial resolution of the micro-Raman spectrometer is 1 μm.

DISCUSSION

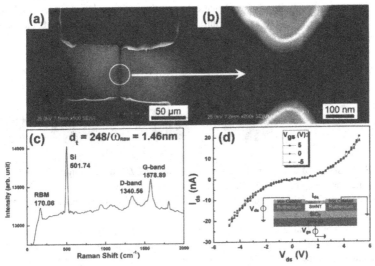

Figure 2. (a) A typical SEM micrograph of the electrode; (b) a typical SEM micrograph the SWNT-bridge structure; (c) a typical Raman spectrum of the SWNTs; and (d) typical I_{ds}-V_{ds} curves of the SWNT-bridge structure with a suspending SWNT channel.

Figure 2(a) shows a typical SEM micrograph of the metallic electrode. Figure 2(b) shows the enlarged image of the circled area in Fig. 2(a), showing the SWNT-bridge prepared via the LCVD process. A SWNT is observed to bridge the electrode pair. The length of the SWNT was determined by the electrode gap distance. The diameter of the SWNT was appeared to be around 12 nm, which was obviously larger than its actual size. A small-amplitude vibration was observed for the SWNT under the electron beam irradiation suggesting a suspending SWNT, which contributed to the appeared large diameter. Figure 2(c) shows a typical Raman spectrum of the SWNT-bridges. Both a sharp RBM peak and a large G-band to D-band ratio clearly indicate the existence of SWNTs [5]. According to the relationship between SWNT diameter (d_t) and RBM frequency (ω_{RBM}) for isolated SWNTs, $\omega_{RBM} = 248/d_t$ [5], the SWNT diameter is estimated to be 1.4 nm. Figure 2(d) shows typical current-voltage curves of the SWNT-bridge structure at different gate biases. The curves show symmetric characteristics which resemble the nonlinear I-V characteristics of SWNTs. The SWNT-bridge structure can be modeled as a metal-semiconductor-metal (M-S-M) structure which consists of two Schottky barriers connected back to back, in series with a semiconductor. The I-V performance of the SWNT-bridge structure shows an increased channel current (I_{ds}) with the increasing channel voltage (V_{ds}), which cannot be explained by the conventional thermionic emission theory which predicts negligible current in a reversely biased Schottky barrier. In this case, the tunneling current under the reversely biased condition is nontrivial and dominates the channel current [6]. The weak gate-voltage (V_g) dependency is ascribed to the weakened gate voltage influence caused by the thick insulating layer of SiO_2 (2 μm) and the air gap (200 nm). Thiss observation strongly supports the formation of the suspended SWNT channel.

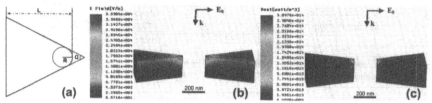

Figure 3. (a) Sketch of the electrode tip used in simulations, L is the tip length, R is the apex radium, and α is the tip angle; (b) HFSS simulation results of electrical field distribution at the electrode tips; and (c) HFSS simulation results of heating distribution at the electrode tips.

It is well known that ultra-sharp metallic tips can be used as optical antennas to localize and enhance optical fields[7]. The optical near-field effects result in an enhanced eddy current and localized heating at the metallic tips. Numerical simulations using the High Frequency Structure Simulator (HFSS, Ansoft) were carried out to evaluate the contribution of the optical near-field effects in the SWNT growth. In the simulation model, the incident laser beam was assumed to be linearly polarized and propagating perpendicularly to the substrate surface. The optical power density of the incident laser beam was assumed to be 15 μW/μm². The electromagnetic frequency was 28.3 THz, equivalent to the CO_2 laser wavelength (10.6 μm). The gap between two tips was set to be 200 nm. The tip radius was assumed to be 20 nm. Figure 3(a) shows the sketch of the electrode tips. Figure 3(b) shows the simulation results of the electrical field distribution around the metallic tips. Significant near field enhancement is observed at the tips, almost tenfold stronger than that of the electrode main body. The enhanced high-frequency electrical field at the tips results in significantly enhanced local eddy currents and yields an enhanced localized heating. Figure 3(c) shows the heat distribution of the electrode tips. Maximum local heating enhancement is observed at the tips, almost one order of magnitude larger than that of the electrode body. Therefore, a significant temperature increase at the tips is expected. An increased temperature at the tips significantly promotes the SWNT growth at the tips compared with the rest part of the electrodes. The thermal simulation results explain the preferential growth of SWNTs at the tips. A temperature increase at metallic tips also benefits the stable SWNT-metal adhesion due to the formation of carbide at high temperature, which promises stable SWNT-bridge structures.

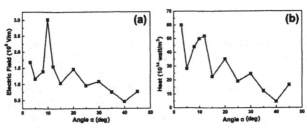

Figure 4. HFSS simulation results demonstrating the influence of tip sharpness on (a) electric near field and (b) localized heating effects.

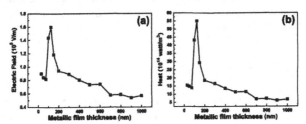

Figure 5. HFSS simulation results demonstrating the influence of metallic film thickness on (a) electric near field, and (b) localized heating effects.

Several parameters influence the electric near field enhancement, including tip geometry, tip length, tip sharpness, film thickness, materials and incident laser wavelength. Noble metals, such as silver and gold, show strong near-field enhancement and are frequently used the tip material. However, due to their high cost, softness, and low melting temperature, these noble metals are not applicable in this study. Ru is chosen as the electrode material for its high melting temperature, stable property and low solubility with Fe and carbon. In the simulations, we designed a Ru triangle tip, as shown in Fig. 3(a), with a fixed length $L = 400$ nm, an opening angle α, and a tip with a fixed radius of curvature $R = 5$ nm. The dependencies of the field enhancement values on the tip angle (α) and film thickness are computed and listed in Figs. 4(a) and 5(a), respectively. Both the angular and film-thickness dependencies exhibit non-monotonic behavior. Maximum electric near filed enhancements are observed at a tip angle of 10° and a film thickness of 125 nm. Figures 4(b) and 5(b) exhibit the dependencies of the heat generation on the tip angle (α) and film thickness, which also show non-monotonic behavior. The maximum heat values are observed at a tip angle of 12° and a film thickness of 125 nm, which are not completely identical with the electric near field enhancement. Electrical resistance, thermal conductivity and thermal capacity have to be taken into consideration for calculating the heat distribution. Mechanisms of the angular and film-thickness dependencies are under investigation.

CONCLUSIONS

In this study, a novel laser-based strategy for *in-situ* SWNT growth was developed using optical near-field effects. Well controlled SWNT-bridge structures with suspending SWNT-channels were fabricated at a significantly reduced substrate temperature. The SWNT growth is tightly confined at the electrode tips. HFSS Simulation analysis shows prominent enhancement of electric near field and localized heat at the tips. Optimal tip sharpness and metallic film thickness for maximum electric near field and heating enhancement are investigated. Further parameters, such as tip length, tip geometry and material, are under investigation to reach maximum field and heating enhancement. Several advantages, including precise position control, low substrate temperature, and reliable SWNT/metal contact, make the LCVD method based on the optical near-field effects a promising technique for future nanoscale fabrication of SWNT-based devices.

ACKNOWLEDGMENTS

This research work was financially supported by National Science Foundation (CMMI 0555884 and ECCS 0621899). The authors would like to thank Profs. N. J. Ianno and R. J. Soukup in the Department of Electrical Engineering, Prof. Y. Zhou in the Center for Biotechnology, Prof. S. H. Liou and Dr. L. P. Yue in the Department of Physics and Astronomy for their technical support and helpful discussion.

REFERENCES

1. H. J. Dai, *Surface Science* **500**, 218 (2002).
2. S. J. Tan, A. R. M. Verschueren, and C. Dekker, *Nature* **393**, 49 (1998).
3. H. J. Dai, *Acc. Chem. Res.* **35**, 1035 (2002).
4. L. Novotny, and S. J. Stranick, *Annu. Rev. Phys. Chem.* **57**, 303 (2006).
5. M. S. Dresselhaus, G. Dresselhaus, R. Saito, and A. Jorio, *Physics Reports-Review Section of Physics Letters* **409**, 47 (2005).
6. Z. Y. Zhang, C. H. Jin, X. L. Liang, Q. Chen, and L. M. Peng, *Applied Physics Letters* **88** 073102 (2006).
7. A. V. Goncharenko, H. C. Chang, and J. K. Wang, *Ultramicroscopy* **107**, 151 (2007).

Mater. Res. Soc. Symp. Proc. Vol. 1142 © 2009 Materials Research Society   1142-JJ10-45

The influence of reaction temperature on the size of the catalyst nano-particle and the carbon nanotube diameter distributions

Enkeleda Dervishi,[1,2*] Zhongrui Li,[2] Fumiya Watanabe,[2] Yang Xu,[2] Viney Saini,[1,2] Alexandru R. Biris,[3] and Alexandru S. Biris[1,2*]

[1]*Applied Science Department, University of Arkansas at Little Rock, 2801 S. University Ave., Little Rock, AR 72204, USA*
[2]*UALR Nanotechnology Center, University of Arkansas at Little Rock, 2801 S. University Ave., Little Rock, AR 72204, USA*
[3]*National Institute for Research and Development of Isotopic and Molecular Technologies, P.O. Box 700, R-400293 Cluj-Napoca, Romania*

* Corresponding authors
Tel: 501-569-3203, Fax: 501-569-8020, Email: exdervishi@ualr.edu
Tel: 501-683-7458, Fax: 501-569-8020, Email: asbiris@ualr.edu

Abstract

Carbon nanotubes (CNTs) with high crystallinity were synthesized on the MgO supported Fe-Co catalyst system using a chemical vapor deposition method. Methane was utilized as a hydrocarbon source and the reaction temperature was varied between 700 to 1000 °C. The influence of the temperature on the size of the metal nano-particles and the morphology of CNTs was systematically studied. The catalyst system Fe:Co:MgO (2.5:2.5:95 wt.%) was exposed to different temperatures and its structural and morphological properties were characterized by microscopy, X-ray diffraction technique and surface area analyzer. It was found that as the temperature increases to 1000 °C, the nanotubes have a wider diameter distribution when compared to the ones grown at lower temperatures. These results correlated well with the microscopy analysis of the catalyst system indicating a variation in the size of the active metal nano-clusters. Furthermore, the morphological changes of the catalysts were directly found to influence the properties of the corresponding CNTs, as obtained from electron microscopy, Raman spectroscopy, and thermogravimetric analysis.

Introduction

Since the discovery of CNTs a lot of work has been done trying to understand the kinetics of the nanotube formation.[1] Synthesis conditions such as reaction temperature, flow rates of the carrier and hydrocarbon gas, hydrocarbon type, catalyst composition and many more, have an enormous effect on the nanotube growth, morphology, and properties. The diameter, chirality and yield of CNTs are very sensitive to the chemical composition of the catalyst system utilized in the growth process.[2] Different transition metals such as Fe, Co, Ni and others are employed as active catalyst materials to decompose the hydrocarbon sources and instigate the nanotube growth.[3] Although, many monometallic catalysts have been successfully used to grow different types of CNTs, bimetallic alloys formed by the transition metals generally provide a more efficient CNT synthesis.[4] In our previous work, we systematically investigated the effects of the Fe-Co interaction on the CNT crystallinity and production efficiency.[5,6]

It is commonly believed that CNT growth is initiated once a carbon monolayer covering a catalyst particle becomes unstable due to the incorporation of additional carbon atoms or thermal vibrations.[7] Obviously, the thermal treatment plays an important role in the nucleation and the carbon nanotube growth. By cautiously selecting the proper synthesis temperature one can control the nanotube diameter distribution. Ago et al. studied the effects of the reaction conditions on the Fe nano-particles supported on porous MgO,[8,12] but so far no thorough studies have been found on the Fe-Co bimetallic catalyst system. The aim of this work is to study the influence of the synthesis temperature on the Fe-Co/MgO catalyst system and the CNT growth properties. Spectroscopy, thermal gravimetrical analysis, microscopy and other techniques were used to analyze the morphological properties of the CNTs and the catalyst system.

Experimental Details

The Fe-Co/MgO catalyst system with a stoichiometric composition of 2.5:2.5:95 wt.%, was prepared by the impregnation method as previously described.[9] First, the weighted amount of metal salts, $Fe(NO_3)_3 \cdot 9H_2O$ and $Co(NO_3)_2 \cdot 6H_2O$, were dissolved separately in ethanol with agitation. Next, MgO with surface area 130 m^2/ g was completely dispersed into 30 ml of ethanol and the metal salt mixtures were added to this MgO solution. Next, the ethanol was evaporated under continuous agitation, and the catalyst system was further dried overnight at 60 °C. Finally, the catalyst was calcinated in air at 500 °C for 2 hours. CNTs were grown by utilizing a Radio Frequency (RF) catalytic Chemical Vapor Deposition (cCVD) method.[10] In this case then RF generator was simply used as an inductive heating source. Approximately 100 mg of the catalyst was uniformly spread into a thin layer on a graphite susceptor and placed in the center of a quartz tube with inner diameter of 1 inch. First, the tube was purged with the carrier gas (Argon) for 10 minutes at 150 ml/min. Next, the RF generator was turned on and when the temperature of the graphite boat reached the desired synthesis temperature, methane (CH_4) was introduced at 40 ml/min for 30 minutes, while the carrier gas was still flowing. The temperature at which the nanotubes were grown was varied between 700 to 1000 °C at 50-degree increments. At the end of each reaction, the system cooled down under the presence of Argon for 10 minutes. Through out this synthesis, Argon was utilized as a carrier gas in order to dilute the hydrocarbon. The as-produced CNTs were purified in one easy step using diluted hydrochloric acid solution and sonication.

Results and Discussions

The catalyst system Fe-Co/MgO was utilized to grow single-wall and double-wall carbon nanotubes at different reaction temperatures. The synthesis temperature was varied between 700-1000 °C, and major differences when it comes to the size controllability of the catalyst nano-particles were noticed when the temperature was set at 800 and 1000 °C. To understand the effect of the reaction temperature on the catalyst morphology, the catalyst systems thermally treated at two different temperatures (800 and 1000 °C) were characterized by microscopy and other techniques. The catalyst system calcinated at 500 °C will be denoted as cat_500, whereas the catalysts thermally treated at 800 and 1000 °C will be referred to as cat_800 and cat_1000 respectively. The scanning transmission electron microscopy (STEM) image of the cat_800, in Fig. 1 (a), shows a large number of metal nano-particles distributed on the MgO support with very similar dimensions. This is supported by the histogram in Fig. 1 (b) which shows a very narrow diameter distribution of metal nano-clusters for cat_800. The diameters of the Fe-Co nano-particles present on the MgO support vary between 1-2.5 nm and a vast number of them have diameters between 1.5-2 nm.

Fig. 1(a) STEM image of the Fe-Co/MgO (2.5:2.5:95 wt.%) thermally treated at 800 °C, (b) the diameter distribution histogram of the metal nano-clusters present on the MgO support with a Gaussian fitting curve, (c) STEM image of the Fe-Co nano-particles present onto the MgO surface for cat_1000, (d) corresponding diameter distribution histogram of the Fe-Co metal nano-clusters with a Gaussian fitting curve.

Fig. 1 (c) shows the STEM image of cat_1000 in which the Fe-Co nano-clusters with many different sizes are present onto the surface of the MgO support. The histogram in Fig. 1 (d) indicates a wide diameter distribution of the Fe-Co nano-particles for the catalyst system thermally treated at 1000 °C. The diameters of the metal nano-clusters vary between 1.5 to 6.5 nm and approximately 27 % of the nano-particles have diameters from 3 to 3.5 nm. As observed from Fig. 1, the Fe-Co nano-clusters have a much wider diameter distribution when the catalyst system Fe-Co/MgO is heated at 1000 °C compared to the one thermally treated at 800 °C. It is believed that at 1000 °C the Fe-Co nano-particles are in molten state when they could break-up into smaller particles or recombine to form larger clusters, generating diameters with a wide variation.[1]

The Langmuir surface area of Fe-Co/MgO catalyst calcinated at 500 °C was measured to be 48.84 m²/g. As the temperature increased from 500 to 800 °C, the surface area of the catalyst decreased to 35.18 m²/g, since at high temperatures the MgO particles agglomerated into larger nano-clusters yielding a lower catalyst surface area. At 800 °C, the MgO support has small nano-pores which provide a better localization for the active metal nano-particles confining them into small nano-clusters. The surface area of cat_1000 decreased, whereas the adsorption pore width increased when compared to cat_800. This increase in the pore width of the cat_1000 might indicate a broadening of the MgO pore size aiding the formation of nano-clusters with a wider diameter distribution. X-ray diffraction (XRD) patterns (not shown here) of the cat_500 and cat_800 indicate the presence of phases characteristic to MgO and the oxides of Fe or Co, but not to Fe-Co alloys. However, the XRD profile of the catalyst thermally treated at 1000 °C, shows an additional peak at 2θ = 44.8 degrees, which is attributed to the Fe-Co alloys. This could be due to the catalyst exposure to high temperature, when it is possible that the Fe-Co nano-particles melt and possibly form alloys.

The TEM image in Fig. 2 (a) of the CNTs grown at 800 °C shows a bundle of single-wall carbon nanotubes (SWCNTs) with very uniform diameters of 3 nm. The nanotubes grown at 800 and 1000°C will be referred to as CNT_800 and CNT_1000 respectively. At 800 °C the nano-particles present in the MgO support have a very narrow size distribution hence yielding SWCNT growth with small and uniform diameters. Fig. 2 (b) shows the SEM image of a very high density of weblike CNTs grown at 800 °C. When the synthesis temperature was increased to 1000 °C, the STEM analysis indicated that the Fe-Co nano-particles have a wide diameter distribution which nucleated the growth of nanotubes with a very broad diameter distribution. Fig. 2 (c) shows the TEM image of SWCNTs with various diameters synthesized at 1000 °C. In addition to SWCNTs, about 10-20 % of double-wall carbon nanotubes (DWCNTs) were also present in the sample synthesized at 1000 °C. This may be due to the large catalytically active nano-particles with diameters between 4.5-6.5 nm which possibly nucleate DWCNT growth.[12] Fig. 2 (d) shows a TEM image of a metal nano-cluster with diameter of 4.5 nm trapped inside a DWCNT grown at 1000 °C. Fig. 2 (e) shows the SEM images of high density CNTs synthesized at 1000 °C.

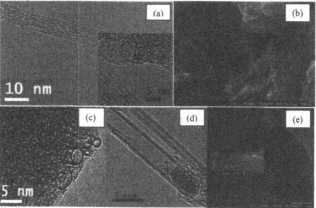

Fig 2 (a) TEM image of the SWCNTs with very uniform diameters synthesized at 800 °C, (b) SEM image of CNT_800, (c) TEM image of the CNT_1000 with various diamters, (d) TEM image of a metal nano-particle trapped inside a DWCNT grown at 1000 °C, (e) the corresponding SEM image of high density nanotubes grown at 1000 °C.

Thermogravimetric analysis (TGA) is a useful technique for characterizing the purity and the thermal stability of CNTs. Fig. 3 (a) shows the weight loss profile curves of the CNT_800 and CNT_1000, revealing that after only one purification step both samples had a purity of higher than 95 %. The inset in Fig. 3 (a) presents the first derivatives of the normalized TGA curves indicating the nanotube combustion temperatures of each sample. The TGA curves and the corresponding differential thermal analysis (DTA) show significant weight losses at 568 °C and 585 °C for the purified CNT_800 and CNT_1000, respectively. The CNTs grown at 1000 °C decomposed at a higher temperature than the ones grown at 800 °C due to the presence of more DWCNTs in the CNT_1000. It has been shown that the thermal decomposition temperature of the CNTs depends on their morphological properties and as the number of their graphitic walls increases so does their combustion temperature.[13,14] Therefore, our TGA analysis correlate well with the TEM findings which indicate a higher percentage of DWCNTs grown at 1000 °C compared to 800 °C. Fig. 3 (b) shows the TGA curves of the as-produced CNTs grown at different temperatures (700–1000 °C, at 50 degree increments). The inset shows the combustion temperatures for the CNTs generated from the first derivative curves of the corresponding TGA curves. The decomposition temperature of the CNTs grown at 1000 °C is higher than the combustion temperature of the CNTs grown at lower temperatures. In addition, the purified CNT_800 and CNT_1000 burn at higher temperatures than the as-produced samples. This is due to the presence of metal nano-particles trapped next to the nanotube walls (in between the bundles) of the un-purified samples, which act as thermal catalysts in the burning process.[15]

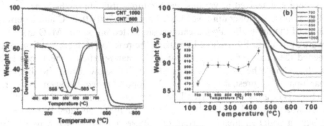

Fig 3 (a) TGA curves of the CNTs grown at 800 and 1000 °C, respectively, (inset) the first derivative curves indicating the corresponding combustion temperatures for CNT_800 and CNT_1000, (b) TGA curves of the as-produced CNTs, (inset) the corresponding combustion temperatures of the CNTs grown between 700–1000 °C, in 50 degree increments.

Raman Spectroscopy was used to analyze the crystallinity and the diameter distribution of CNTs grown at various reaction temperatures. The vibrational modes observed in the nanotube Raman spectra are the Radial Breathing Mode (RBM), the D band, G band and the 2D band.[16] Theoretical calculations of SWCNTs have shown that tube diameter d and the radial mode frequency ω_{RBM} exhibits the following straightforward relationship: $\omega_{RBM}(cm^{-1}) = \dfrac{\alpha}{d(nm)} + a$ where $\alpha = 234\ cm^{-1} \cdot nm$ and $a = 10$ are constants that depend upon the excitation energy and bundle sizes.[17,18]

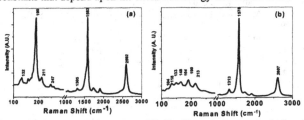

Fig. 4 (a) Raman Spectrum of the CNTs grown at 800 °C utilizing methane as the hydrocarbon source, (b) Raman Spectrum of the CNTs grown at 1000 °C collected with a 633 nm laser excitation.

Fig. 4 (a) shows the Raman spectrum of the SWCNTs grown at 800 °C collected with the 633 nm laser excitation. These SWCNTs have a very intense peak in the RBM region at 190 cm^{-1} which correspond to the nanotube diameter of 1.3 nm. Fig. 4 (b) shows the Raman spectrum of the CNTs grown at 1000 °C indicating a large number of peaks in the RBM region corresponding to a broad nanotube diameter distribution. The dominant diameter distribution of the CNT_800 ranges from 0.98 to 1.91, whereas the nanotubes grown at 1000 °C have a wider diameter distribution between 1.15 to 2.49 nm. The

Raman spectra of CNT_1000 have a larger number of RBM peaks corresponding to different diameters of SW or DWCNTs.

Conclusions

By using the highly selective Fe-Co/MgO catalytic method, high quality SWCNTs were synthesized at various temperatures utilizing methane as a hydrocarbon source. The synthesis temperature was varied between 700-1000 °C and its effects on the morphology of the bimetallic Fe-Co catalyst supported on porous MgO and on the CNT growth were systematically studied. Increasing the reaction temperature from 800 to 1000 °C broadened the diameter distribution of Fe-Co nano-particles and hence affected the nanotube diameters. The STEM analysis of Cat_800 showed that the diameters of the Fe-Co nano-particles present on the MgO support varied between 1-2.5 nm hence yielding SWCNT growth with small and uniform diameters. When the synthesis temperature was increased to 1000 °C, the STEM analysis indicated that the Fe-Co nano-particles have a wider diameter distribution (between 1.5-6.5 nm), which nucleated the growth of nanotubes with many different diameters. Therefore, an increase in synthesis temperature leads to SWCNT growth with a wider diameter distribution as well as a small amount of DWCNTs. These results are also supported by thermogravimetric and Raman spectroscopy analysis. These finding provide a simple method to optimize the diameter distribution of SWCNTs by choosing a proper reaction temperature.

References

1 Iijima S., *Nature* (London, United Kingdom) **1991**, 354(6348), 56-8.

2 Ch. Laurent, E. Flahout, A. Peigney, A. Rousset, *New J. Chem.* **1998**, 1229-1237.

3 S. Seraphin et al., *Chem. Phys. Lett.* **1994**, 217, 191.

4 T. Guo, P. Nikolaev, A.Thess, D.T. Colbert, R. E. Smalley, *Chem. Phys. Lett.* **1995**, 243, 49.

5 Zh. Li, E. Dervishi, Y. Xu, X. Ma, V. Saini, A. S. Biris, R. Little, A. R. Biris, D. Lupu, *Journal of Chemical Physics* **2008**, 129(7), 074712/1-074712/6.

6 E. Dervishi, Z. Li, A. R. Biris, D. Lupu, I. E. Pavel, S. Trigwell, and A. S. Biris, *Chem. Mater.* **2007**, 19(2), 179.

7 J.-C. Charlier, S. Iijima, *Topics in Applied Physics* **2001**, 80, 55-80.

8 H. Ago, K. Nakamura, N. Uebara, and M. Tsuji, *J. Phys. Chem. B* **2004**, 108, 18908-18915.

9 S.C. Lyu, B.C. Liu, S.H. Lee, C.Y. Park, H.K. Kang, C.W. Yang, C.J. Lee, *J. Phys. Chem. B* **2004**, 108, 1613-1616

10 A. R. Biris, A. S. Biris, E. Dervishi, D. Lupu, S. Trigwell, Z. Rahman, P. Marginean, *Chemical Physics Letters* **2006**, 429, pp. 204-208.

11 D. Feng, R. Arne and B. Kim, *Carbon* **2005**, 43, (10), 2215-2217.

12 H. Ago, K. Nakamura, Sh. Imamura, and M. Tsuji, *Chem. Phys. Lett.* **2004**, 391, 308-313.

13 S.B. M. Gregg, S. V. Kenneth, *J. Phys. Chem. B* **2006**, 110, 1179.

14 S. Arepalli, P. Nikolaev, O. Gorelik, V.G. Hadjiev, W. Holmes, B. Files, L. Yowell, *Carbon* **2004**, 42, 1783.

15 A. C. Dillon, T. Gennett, M. J. Kim, J. L. Alleman, A. Parilla, M.J. Heben, *Adv. Mater.* **1999**, 11, 1354,

16 M.S. Dresselhaus, G. Dresselhaus, A. Jorio, A.G. Souza Filho, R. Saito, *Carbon* **2002**, 40, 2043-2061.

17 H. Kuzmany, W. Plank, M. Hulman, Ch. Kramberger, A. Gru¨neis, Th. Pichler *Euro Phys J. B.* **2001**, 22, 307-20.

18 A. M. Rao, J. Chen, E. Richter, U. Schlecht, P. C. Eklund, R. C. Haddon, U. D. Venkateswaran, Y.-K. Kwon, and D. Tománek, *Physical Review Letters* **2001**, 86, 17, 23.

Optical and Thermal Properties

Mater. Res. Soc. Symp. Proc. Vol. 1142 © 2009 Materials Research Society

Carbon Nanotubes for Heat Management Systems: Black Body Radiation and Quenching of Phonon Modes in MWNT Bundles

Ali E. Aliev [*],[1] Gautam K. Hemani,[1] Edward M. Silverman,[2] Ray H. Baughman[1].
[1] NanoTech Institute, University of Texas at Dallas, Richardson, TX, 75083, USA
[2] Northrop Grumman Space Technology, Redondo Beach, CA 90278, USA

ABSTRACT

The extremely high predicted and observed thermal conductivity of individual carbon nanotubes has not yet been achieved for macroscopic nanotube assemblies. Thermal resistances at tube-tube interconnections and tube/electrode interfaces are considered main obstacles for effective electronic and heat transport. Here we show that, even for infinitely long, perfect nanotubes with well designed nanotube/ electrode interfaces, excessive radial heat radiation from nanotube surfaces and quenching of phonon modes in high density bundles are additional processes that substantially reduce thermal transport along the nanotubes. Self-heating 3ω and laser flash techniques were employed to determine the peculiarities of anisotropic heat flow and the thermal conductivity of a single MWNT, bundled MWNTs and aligned freestanding MWNT sheets. The thermal conductivities of individual MWNTs grown by CVD methods and normalized the density of graphite (κ=600 W/m·K) are much lower than theoretically predicted. Coupling within MWNT bundles decreases this thermal conductivity to 150 W/m·K. Further decrease of effective thermal conductivity in MWNT sheets to 50 W/m·K comes from sheet imperfections like dangling fiber ends, loops and misalignment of nanotubes. Optimal structures for enhancing thermal conductivity are discussed.

INTRODUCTION

Bulk samples of well aligned carbon nanotubes represent a fascinating material, with highly anisotropic electrical and thermal transport properties. Theoretical predictions yield an extremely high thermal conductivity for individual carbon nanotubes (CNT), κ=6600 W/m·K [1], and suggest their possible application for heat management systems [2]. For macroscopic devices, one must assemble a vast number of nanotubes in order to obtain a desired conductance. Then interaction between confined nanotubes in bundles can partially decrease their rotational and vibrational freedom and lead to quenching of phonon modes.

We will restrict our discussion to the use of multi-walled carbon nanotube (MWNT) structures as a preferred choice for phonon propagation: As the nanotube diameter increases, more optical phonon modes are excited which contribute to the heat flow. The higher strain fields in inner shells of MWNT can result in higher thermal conductivity compared to single wall carbon nanotubes [3]. Moreover, it is naturally to think that intrinsic defects (vacancies, or conformations) should have much more severe effects in SWNTs (one-dimensional conductor) than in MWNTs (two-dimensional conductor). The neighboring shells in MWNT provide effective additional channels for phonons to bypass the defective sites. Additionally, the outermost shell protects the inner shells from the environment.

Recently we have developed a new dry-state technique to produce highly-oriented, freestanding MWNT sheets that appear to be very attractive for thermal transport applications [4]. High alignment, the two dimensional structure of the MWNT sheets and extensive tube-tube overlap of individual tubes are potentially promising features for effective heat transport. By

---

[*] Corresponding author. Tel.: +1 972-883-6543; fax: +1 972-883-6529. E-mail address: Ali.Aliev@utdallas.edu.

measuring thermal conductivity versus electrode separation length and temperature we have experimentally demonstrated that heat transport in long free-standing samples is dominated by surface radiation, so that little heat energy is transferred by phonons for distances > 2 mm [5,6]. The thermal conductivity (κ) and the thermal diffusivity (α) of highly aligned MWNT sheet (measured for short distances, L<2 mm) are relatively low (50±5 W/m·K and 45±5 mm^2/s, respectively) and mostly deteriorated by the low quality of CVD grown carbon nanotubes (intrinsic defects of individual nanotubes), phonon scattering in bundles and imperfect alignment.

If extremely large surface area and high emissivity of carbon nanotubes cause the high surface radiation, then a solution for this problem is a bulk dense structure where the outermost nanotubes will screen radiation from the interior. However, coupling between nanotubes in bundles can suppress the phonon modes and decreases the transport abilities of individual MWNTs. The thermal conductivity (600±100 W/m·K) of individual CVD grown MWNTs (obtained by 3ω method) is lower than theoretically predicted. Coupling within nanotube bundles and poor impedance matching at tube-sink interfaces decreases the thermal conductivity to 150±20 W/m·K. The further decrease of effective thermal conductivity in MWNT sheets to 50±5 W/m·K comes from sheet imperfections like fiber ends, loops and misalignment of tubes.

EXPERIMENTAL DETAILS

Highly oriented transparent nanotube sheets and yarns were drawn from the sidewall of a 300-350 μm tall MWNT forest, which was synthesized by a catalytic chemical vapor deposition (CVD) method [4]. To study the quenching of phonon modes a four-probe electrodes assembly was fabricated on non-doped, thermally oxidized silicon and glass substrates (10x10 mm^2) using UV lithography. Two sets of patterned substrates with different electrode separation (2, 4, 6 and 10, 20, 30 μm) have been prepared. A tiny MWNT ribbon pilled off from the suspended sheet or MWNT forest and attached to the four-probe electrodes comprises a numbers of bundles with large diameter (100-150 nm) interwoven by small diameter bundles or single MWNTs. To choose the desired bundle or single MWNT we used Zyvex nano-manipulator (Zyvex Inc., nano100A) equipped with four horizontal fingers ended by tungsten tips (100 nm on the edge).

The quasi 2D structure of the MWNT sheet forms almost a perfect interface with smooth gold electrodes. However, the round shape of large bundles prevents good electrical and thermal contact with the sink. Further improvement of MWNT/electrode interface was obtained by deposition of platinum spots or elongated strip using FIB nano200. The samples with low numbers of MWNTs in bundles have high resistance (R>0.5 MΩ) and are very sensitive to the static potential, so they are difficult to handle. To test the quality of MWNT/electrode contacts, the resistance of left (20 μm) and right (30 μm) end segments (compared with the resistance of middle electrode separation,10 μm) were characterized by 2 and 4-probe methods using a Keithley 2425-C Source Meter and a low probe current, I<10 μA.

Thermal transport measurements were performed using self-heating 3ω and laser flash techniques. The diffusive nature of electronic conductivity of MWNT suggests the use of the self-heating 3ω technique to determine the thermal transport along the one-dimensional (1D) conductor. The details of used techniques can be find elsewhere [5, 6]. To minimize the high $U_{1\omega}/U_{3\omega}$ ratio and phase uncertainty of suspended end segments, we designed two-probe reading with the $U_{1\omega}$ signal compensated using a Wheatstone bridge. All measurements were done under high vacuum (P=0.1 mbar) in a Janis Research VPF-475 cryostat to reduce the radial heat losses through gas convection. To minimize static radial heat dissipation from the specimen we used a simple heat shield using aluminum foil.

166

RESULTS

Black-body radiation. Measurement of apparent κ for a free-standing MWNT sheet as a function of electrode separation using the 3ω method revealed a quadratic decrease of the third harmonic signal and consequently an increase in the derived thermal conductivity for longer specimens (Fig.1A). Below 2 mm sample length, the apparent thermal conductivity saturates at a κ =50±5 W/m·K, which will be shown to correspond to the real thermal conductivity.

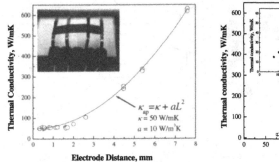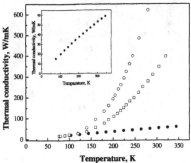

Fig. 1. (Left) Distance dependence of apparent κ (open circles) for a MWNT sheet measured using the 3ω method. The solid line is the calculated κ obtained using Eq.(1). Fitting parameters are shown in right bottom inset. The top inset shows the 4-probe geometry of the sample. (Right) The temperature dependence of κ for three different lengths: 7.6 mm (red open circles), 5.4 mm (open squares) and 0.37 mm (solid circles). The inset provides an expanded plot for the L=0.37 mm sample.

What causes the rise of apparent κ with increased length? For a MWNT sheet with extremely large surface area the apparent thermal conductivity can be higher than the actual value, $\kappa_{ap}= \kappa + \kappa_{loss}$ due to radial heat loss through radiation [6],

$$\kappa_{loss} = \frac{16\varepsilon\sigma T_o^3 L^2}{\pi^2 d}, \qquad (1)$$

where σ = 5.67·10^{-8} W/m^{2}K^4 is the Stefan-Boltzmann constant, ε is the emissivity, d is the average diameter of individual 1D heat channels, T_o =295 K is the environmental temperature. For porous carbon material we can use the black body emissivity, $\varepsilon=1$. The solid line in Fig.1A plotted using Eq. (1) shows very good agreement with the measured experimental data above L=2 mm. The fitting parameters shown in insets yield the diameter of a cylindrical rods, d=140 nm, which determined the radiation surface area of specimens. This value is surprisingly close to the bundle diameter we found from AFM analysis [5].

Our hypothesis that the measured thermal conductivity is caused by radial heat losses was tested by comparing the temperature dependence of κ for long and short samples. At low temperatures the black body radiation should be significantly reduced and one can measure the intrinsic phonon transport parameters. Fig.1B shows that the κ of all samples are comparable below T=150 K. Above of 150 K, the measured κ of long samples (7.6 and 5.4 mm) are substantially higher than that of the shortest sample (L=0.37 mm). Here it is worth noting that for all of the above experiments the temperature modulation of the suspended part of the sample did not exceed 5 K. Therefore, we believe that for MWNT sheets pulled out from 300 μm forest the radiation losses dominate for sheet lengths above 2 mm. The radial heat current was an order of magnitude higher than the axial current at a sheet length of ~7.6 mm.

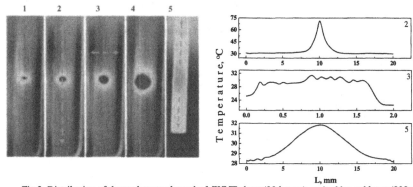

Fig.2. Distribution of thermal spots along the MWNT sheet (20 layers) excited by red laser (830 nm). The power of the excitation beam is increased from left to right in the picture columns (10; 20; 30 and 50 mW for pictures 1-4, respectively). For comparison, the picture in column 5 shows the temperature distribution in a Ge strip excited with power 80 mW. The temperature distribution along the green arrows on the left is shown in plots on the right.

The temperature distribution along the MWNT sheets (taken using a MobIR M4 thermal camera, Wuhan Infrared Tech. Co.) confirms the high radiation losses from the surface. The heat spot created by IR laser (830 nm) is localized around the excited beam and expanded only to 1-2 mm along the sheet, whereas for Ge strip of 0.5x2x20 mm^3 (right column) with almost the same κ (58 W/m·K) the temperature distribution is much more uniform. Due to the high reflectivity of Ge strip the incident beam power was increased to 80 mW. Green arrows on the left columns 2, 3, and 5 mark the line for analysis of temperature profile shown in the plots on the right in Fig. 2.

The white elongated lines radiating from the red laser spot are heat conduction channels arising from dense clusters of large diameter MWNT bundles. Despite the high radiation from the more homogeneously distributed MWNTs in other regions of the sheet, the large bundles conduct thermal energy over long distances. The diagram 3 shows the temperature profile across the sheet taken on the distance of 5 mm apart from the laser spot. The outer layer of a bundle shields the black-body radiation from the inner part and provides the long-range heat channel. However, the bundling of individual tubes should quench some phonon modes that contribute to the κ and provide additional phonon scattering in the physically cross-linked structure. Below we experimentally study phonon scattering in bundles having different numbers of tubes in a bundle.

Phonon Scattering in Bundles. Fig.3 shows that tube-tube coupling substantially decreases room-temperature thermal conductivity of MWNT bundles by a factor of 4 relative to the single tube. The thermal conductivity of single MWNT measured for 10 μm electrode separation and normalized to the ideal high density structure (the inner void space was subtracted) with van der Waals spacing between shells is 600±100 W/m·K. This value is much lower than theoretically predicted κ=6600 W/mK, [1] and that experimentally confirmed for individual nanotubes in [2], κ=3000 W/mK. It is well known that MWNTs grown by CVD method are more defective than those grown by arc-discharge or laser-ablation methods. The coupling nanotube bundles exponentially decreases the thermal conductivity. The decrease of κ is saturated at 120±15 W/m·K when the diameter of MWNT bundles reaches 120-150 nm. The obtained κ of those bundles is consistent with the thermal conductivity of MWNT sheet in which nanotubes in well aligned bundles (with average diameter of 140 nm) are interwoven with poorly oriented nanotubes that laterally interconnect these bundles. The electrical resistance of single MWNT (D=10 nm, d=5 nm) measured by 4-probe method was 56±5 kΩ per 1 μm length.

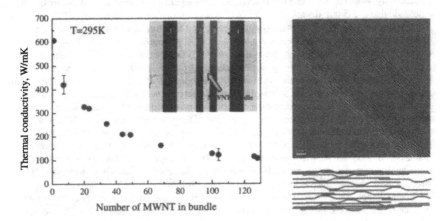

Fig.3. Left: Thermal conductivity of MWNT measured by 3ω method at room temperature as a function of increasing number of tubes in bundles. The inset shows an SEM image of the 4-probe gold patterned substrate with attached MWNT bundle. The electrode separation are: 1-2=30 μm; 2-3=10 μm; 3-4=30 μm. Right top: TEM image of ~10 nm MWNT comprising 7 shells. Right bottom: An optimal structure of MWNT assembly for heat management, which preserves the vibrational freedom of individual nanotubes and shields the heat radiation due to dens outermost layer.

The low-frequency $\omega(k)$ band diagram of MWNTs is very similar to the graphite and has four acoustic branches, which originate from the Γ point of the Brillouin zone with zero frequency ω at the zero wave vector, k. In the order of increasing frequency, these are doubly degenerated transverse (shear) modes involving out-of-plane motion, in-plane twisting modes (bond bending), and in-plane longitudinal modes (bond stretching). The twisting and longitudinal modes involve only movement of atoms tangential to the tube surface and thus are expected to couple very weakly to the other tubes in bundle. The lowest frequency optical branch at k=0 corresponds to the squeezing modes associated with a change of the circular tube cross section to elliptical. Since the bending, transverse and squeezing modes involve out-of-plane motion, they are suppressed by surrounding tubes.

Berber et al mention in [1] that strong tube-tube coupling can decrease the high-temperature thermal conductivity of SWNT bundles by an order of magnitude relative to isolated tubes. The same occurs for graphene layers when they are stacked in graphite - the interlayer interactions quench the thermal conductivity of this system by nearly one order of magnitude. Moreover the van der Waals forces between adjacent SWNTs can deform them substantially, destroying cylindrical symmetry [7]. The experimentally obtained flattening of the tubes along the contact region in very similar MWNTs [8] can substantially reinforce the tube-tube coupling and cause an additional suppression of twisting and shear phonon modes.

Let suppose that we have achieved a 1 m long carbon nanotube with κ=3000 W/mK. Radial radiation from the surface will dissipate heat energy four order of magnitude more than heat propagating along the tube - κ_{loss}=2.45x10^8 W/mK for a 10 nm diameter nanotube. For a 1 cm long MWNT the radial losses are still one order of magnitude higher than the axial thermal conductivity. Both radiation and thermal conductivity equally contribute to heat dissipation from a hot side that is 3.5 mm distant.

The key question is "What is the optimal length and structure of nanotubes assembly to obtain highest thermal conductivity for a long heat management system. As we have seen above, to reduce the radial radiation from the surface of MWNTs we need to assemble the nanotubes in a dense structure where outer shell prevents radiation from the interior. However, to reduce the inter-tube scattering in bundles the tubes should touch each other only over short distances, while still providing sufficient overlap to transfer thermal energy between nanotubes. Taking account for the strong van der Waals interaction between nanotubes and the 340 times lower κ in the perpendicular direction, using computer simulation of thermal resistance in 7 shell MWNTs, we estimated the optimal overlap length to be L~35 μm. Now, if we want to preserve the high individual nanotube κ (at least κ >400 W/mK) the nanotubes should touch each other over less than 2-3% of total length, which provides a total MWNT length of 1.2-1.8 mm (assuming the same quality of nanotubes as presently investigated). The schematic sketch of preferable 2D structure with short overlap is shown in Fig. 3 (right bottom).

In conclusion, we show that bundling decreases thermal conductivity, while at the same time decreasing the ratio thermal emission to phononic transport along the nanotubes. The quenching of phonon modes in bundles, reinforced by radial deformation of carbon nanotubes by van der Waals forces, substantially reduces the transport abilities inherent to individual carbon nanotubes.

ACKNOWLEDGEMENTS

We gratefully acknowledge M. Zhang and S. Fang for providing us with the MWNT sheets. Research supported by the Strategic Partnership for Research in Nanotechnology (SPRING) via the AFOSR, NSF Grant DMI-0609115, and a Northrop Grumman Space Technology grant.

REFERENCES

1. S. Berber, Y-K. Kwon, D. Tomanek. *Phys. Rev. Lett.* **84**, 4613 (2000).
2. P. M. Ajayan, O. Z. Zhou, Application of Carbon Nanotubes. *Carbon Nanotubes*, ed. M. S. Dresselhaus, G. Dresselhaus, Ph. Avouris, (Springer-Verlag, 2001) pp.391–425.
3. T. Kim, M. *et al*, *Phys. Rev. B* **76**, 155424 (2007).
4. M. Zhang, S. Fang, A. A. Zakhidov, S. B. Lee, A. E. Aliev, C. D. Williams, K. R. Atkinson, R. H. Baughman, *Science* **309**,1215 (2005).
5. A. E. Aliev, C. Guthy, M. Zhang, S. Fang. A. A. Zakhidov, J. E. Fischer, R. H. Baughman, *Carbon* **45**, 2880 (2007).
6. L. Lu, W. Yi, D. L. Zhang, *Review of Scientific Instruments* **72**, 2996 (2001).
7. R. S. Ruoff, J. Tersoff, D/ C. Lorents, S. Subramoney, B. Chan. *Nature* **364**, 514 (1993).
8. Ph. Avouris, T. Hertel, R. Martel, T. Schmidt, H. R. Shea, R. E. Walkut, *Applied Surface Science* **141**, 201 (1999).

Nanowires and Nanotubes:
Electrical, Optical and Thermal Properties

Mater. Res. Soc. Symp. Proc. Vol. 1142 © 2009 Materials Research Society   1142-JJ15-14

Gate Controlled Negative Differential Resistance and Photoconductivity Enhancement in Carbon Nanotube Addressable Intra-Connects

Seon Woo Lee[1], Slava Rotkin[2], Andrei Sirenko[3] ,Daniel Lopez[4], Avi Kornblit[4] and Haim Grebel[1]

[1]Electronic Imaging Center at NJIT and the Department of Electrical and Computer Engineering, New Jersey Institute of Technology, Newark, NJ, 07102, U.S.A
[2]Department of Physics, Lehigh University, Bethlehem, PA, 18015
[3]Department of Physics, New Jersey Institute of Technology, Newark, NJ, 07102, U.S.A
[4]New Jersey Nanotechnology Consortium (NJNC), Alcatel-Lucent Technologies Bell Labs, Murray Hill, NJ, 07974, U.S.A

ABSTRACT

Field effect transistors were fabricated using carbon nanotube (CNT) intra-connnects. The intra-connects – individual tube or a small bundle of tubes spanning across the planar electrodes – were grown by using chemical vapor deposition (CVD) precisely between very sharp metal tips on the pre-fabricated electrodes. Gate-controlled N-shaped negative differential resistance (NDR) has been demonstrated. Enhanced differential photoconductance, which was associated with NDR was observed, as well.

INTRODUCTION

Single-wall carbon nanotube (SWCNT) intra-connects (bridges spanning across two planar electrodes, Figure 1) have gained much interest from basic research and potential applications points of view [1-6]. Despite the key role of this relatively simple structures, only a few experimental studies have been devoted to its related negative differential resistance effect [7-10]. The lack of many experiments might be attributed to the lack of a controlled fabrication method. Here we address this issue by devising a new processing method: in some cases, more that 50% successful bridges were fabricated between pre-fabricated set of electrodes. We achieved this value by using sharp electrode tips as seed catalysts to initiate tube growth from tip to tip. All of the intra-connects fabricated in this way were SWCNT. We used this method to study the characteristics of gated structures with individual tubes or small bundles.

EXPERIMENT

SWCNT intra-connects were grown and self-aligned between pre-fabricated electrodes by Chemical Vapor Deposition (CVD) process [11,12]. For the present study we replaced the CO precursor gas by methane/hydrogen mixture at the elevated temperature [13]. The schematic of a field effect transistor (FET) made of individually addressed CNT intra-connect is shown in figure 1b. The silicon surface was oxidized to a thickness of 20 nm prior to the electrode deposition. The two metal electrodes were used as source and drain electrodes, respectively. The silicon substrate was used as a back gate electrode. The growth yield of such technique is quite high (ranging between 30-50%) yet, the exact reason for its success is not clear. It has been postulated in the past that the growth of a single channel is advanced by the formation of suspended graphene between the electrodes, which quickly rolls into a carbon nanotube [14].

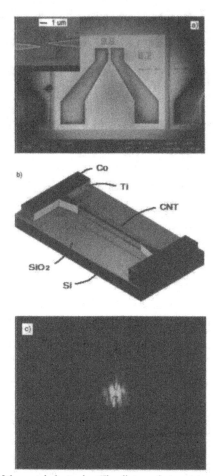

Figure 1. (a) SEM image of the metal electrodes. The distance between two sharp tips was 1 μm. Individually addressable CNT intra-connect was grown directly from tip to tip by CVD. (b) Schematic diagram of our CNT FET. Co was deposited as a catalyst on top of Ti electrode to initiate CNT intra-connect growth. Si substrate was used as a back gate. (c) Laser spot between electrodes

Raman spectroscopy was used in order to determine the type of SWCNT channel between electrodes: metallic or, semiconductor. Raman spectra were measured using the 514.5 nm line of an ion Ar^+ laser, a single grating spectrometer equipped with two notch filters for the laser line, and a N_2 cooled CCD detector. The system was equipped with a x50 microscope to aid focusing of the laser light between the sharp tips of the electrodes (Figure 1c). Broadband light illumination was made with a white-light tungsten source with high energy cut-off at 3 eV.

RESULTS AND DISCUSSION

Figure 2 shows SEM, Ids-Vds characteristics and Raman spectra of CNT intra-connect. The nonlinear I-V curve at zero gate voltage indicates a presence of a barrier between the tube and metal contacts. The asymmetry in the curve points to one dominant contact barrier over the other (otherwise, the curve would have been symmetric). Raman spectra are shown in Figure 2c. The low frequency Raman spectra exhibited a narrow single peak (5 cm^{-1} wide and limited only by the system resolution) for the radial breathing mode (RBM) at 191.9 cm^{-1}. Judged by the narrow Raman RBM spectra, the channel is probably made of a single tube or, made of a small bundle of tubes. The RBM peak corresponds to a metallic tube (12,6) with an average diameter of 1.252 nm [15]. We note however, that if the tubes are lying on the substrate, a spectral distortion may occur and the detrmination of the chirality (m,n) is obscured [16]. In fact, based on high-frequency analysis, in which the difference between the G$^+$ and G$^-$ lines is taken into account, the tube channel may be identified as semiconductive [17]. Such aspects require further investigation.

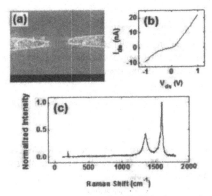

Figure 2. (a) SEM image of CNT intra-connect (b) Ids-Vds characteristic. (c) Raman spectra with RBM at 191.9 cm^{-1}.

Current-voltage Ids-Vds characteristics were measured for various Vgs from -10 V to +5 V. Negative differential resistance (NDR) was found in the Ids-Vgs curves for gate bias values in the region of -3>Vgs>-6 V (Figure 3a). The NDR peak was shifted to the negative side as the source-drain voltage increased from Vds=0 to 0.75 V. Otherwise, the intra-connects exhibited characteristics of an ordinary p-type channel. The current (and the conductivity of the channel) was relatively low. It weakly depends on gate voltage and one may attribute such behavior to mostly the Schottky point of interface.

The experiments were repeated under white light illumination (Figure 3b). The light increased the carrier density in the channel, but not in the metal electrodes. It allowed us to study the effective doping of the channel without affecting the work function of the SWCNT/metal contact. The overall channel conductance increased under the light irradiation. Under illumination, the devices became more stable, as well. While the origin of such behavior is unknown at this point, one may postulate that it may be attributed to screening based on the similarity between the trends in Figure 3a and 3b. The optical effect is relatively small due to the impeding effect of the contact barrier. Moreover, since photo-excited carriers are composed of both types (holes and electrons), n-type characteristics at large positive gate voltages ought to be exhibited as indeed indicated in Figure 3b. One may note the increase in the peak-to-valley current ratio under illumination (Figure 3b). This is puzzling and requires further investigations.

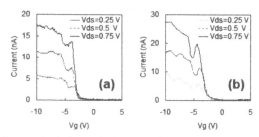

Figure 3. Electrical properties before and after white light illumination. (a) Ids-Vgs characteristics. Negative differential resistance was observed for Vgs between -2 and -6. (b) Ids - Vgs characteristics after white light illumination.

CONCLUSIONS

We have investigated contact properties between a SWCNT intra-connect and metal electrodes in a controlled layout settings. Gate related negative differential resistance, associated with enhanced photoconductance, was observed.

REFERENCES

1. X. Wang, L. Zhang, Y. Lu, H. Dai, Y. K. Kato and E. Pop, *Appl Phys. Letts.* **91**, 261102 (2007).
2. S. J. Kang, C. Kocabas, T. Ozel, M. Shim, N. Pimparkar, M. A. Alam, S. V. Rotkin and J. A. Rogers, *Nature Nanotechnology* **2**, 230 - 236 (2007).
3. S. J. Tans, M. H. Devoret, H. Dai, A. Thess, R. E. Smalley, L. J. Geerligs and C. Dekker, *Nature* **386**, 474-477 (1997).
4. W. Lu and C.M. Lieber, *Nature Mater.* **6**, 841-850 (2007).
5. K. Bosnick, N. Gabor and P. L. McEuen, *Appl. Phys. Lett.* **89**, 163121 (2006).
6. Z. Chen, J. Appenzeller, Y. M. Lin, J. Sippel-Oakley, A. G. Rinzler, J. Tang, S. J. Wind, P. M. Solomon and P. Avouris, *Science* **311**, 1735 (2006).

7. C. Zhou, J. Kong, E. Yenilmez and H. Dai, *Science* **290**, 1552 (2000).

8. A. Iliea, S. Egger, S. Friedrichs, D. J. Kang and M. L. H. Green, *Appl. Phys. Letts.* **91**, 253124 (2007).

9. X. F. Li, K. Q. Chen, L. Wang, M. Q. Long, B. S. Zou and Z. Shuai, *Appl. Phys. Letts.* **91**, 133511 (2007).

10. A. Javey, J. Guo, Q. Wang, M. Lundstrom and H. Dai, *Nature* **424**, 654 (2003).

11. D. Katz, D. Lopez, A. Kornblit and H. Grebel, *J. Nanotechnology and Nanostructures* **8**, 1-5 (2008).

12. D. Katz, S-W. Lee, D. Lopez, A. Kornblit and H. Grebel, *J. Vac Sys. and Tech. B* **25**, 1191 (2007).

13. J. Kong, H. T. Soh, A. Cassell, C. F. Quate and H. Dai, *Nature* **395**, 878 (1998).

14. H. Grebel, private communication.

15. R. Krupke, F. Hennrich, H. V. Lohneysen and M. M. Kappes, *Science* **301**, 344-347 (2003).

16. Yingying Zhang, Hyungbin Son, Jin Zhang, Mildred S. Dresselhaus, Jing Kong, and Zhongfan Liu, *J. Phys. Chem. C*, 111 (5), 1983-1987 (2007)

17. A. Jorio, A. G. Souza Filho, G. Dresselhaus, M. S. Dresselhaus, A. K. Swan, M. S. Unlu, B. B. Goldberg, M. A. Pimenta, J. H. Hafner, C. M. Lieber, and R. Saito, *Phys. Rev B* **65**, 155412 (2002).

Mater. Res. Soc. Symp. Proc. Vol. 1142 © 2009 Materials Research Society 1142-JJ15-25

Electronic, structural and transport properties of nicotinamide and ascorbic acid molecules interacting with carbon nanotubes

Vivian M. de Menezes[1], Ivana Zanella [2], Ronaldo Mota[1], Alexandre R. Rocha[3,4], Adalberto Fazzio[3,4], and Solange B. Fagan[2]

[1]Departamento de Física, Universidade Federal de Santa Maria, UFSM, 97105-900, Santa Maria, RS, Brazil
[2]Área de Ciências Naturais e Tecnológicas, Centro Universitário Franciscano, UNIFRA, 97010-032, Santa Maria, RS, Brazil
[3]Instituto de Física, Universidade de São Paulo, USP, Caixa Postal 66318, 05315-970, São Paulo, SP, Brazil
[4]Centro de Ciências Naturais e Humanas, Universidade Federal do ABC, 09210-170, Santo André, SP, Brazil

ABSTRACT

Ab initio simulations of the nicotinamide (vitamin B3) and ascorbic acid (vitamin C) molecules adsorbed on single-walled carbon nanotubes are performed based on density functional theory. Using a combination with non-equilibrium Green's functions methods, the electronic transport properties of these molecules adsorbed onto semiconducting nanotubes are also studied. The adsorptions of these molecules on the nanotube surface are observed to depend strongly on the functionalization of the adsorbed species. It is demonstrated that when the functionalized vitamins are adsorbed on the nanotube via a strongly covalent bond, significant changes on the electronic transport properties of the nanotubes are verified. In all cases, significant reductions of the total transmissions, both at the valence and conduction bands, of the nanotubes are observed. In some cases, a sharp Fano-type resonance appears, indicating a weak coupling between the sharp states of the molecule and the block states belonging to the nanotube. Hence, it is remarkable to observe that carbon nanotubes adsorbing molecules could result in promising vitamin carriers in both, pristine or functionalized, forms.

INTRODUCTION

Carbon nanotubes (CNTs) have attracted great interest of the scientific community due to their prominent features since Iijima's paper in 1991 [1]. These structures, associated with pharmaceutical compounds, can develop promising systems for drug dissemination or biological sensors. However, to successfully use the single-walled carbon nanotubes (SWNTs) in drug

179

delivery, there are some points that must be taken into account, for example, the biological compatibilities of the carbon nanostructures [2] and the type of the chemical binding between the drugs and these carbon structures [3].

The nicotinamide (VTB) molecule is necessary for good blood circulation and skin health, meanwhile, ascorbic acid (VTC) is important for the immune response system, acting as triglycerides and cholesterol reducer, and helping to decrease side effects of particular drugs and pollutants. Both vitamins molecules are part of many other chemical complexes and soluble in water, but present certain instability that can be controlled by the association with other chemical species such as, for example, SWNTs, which are highly stable molecules.

In this context, this work studies structural and electronic properties of the interactions between semiconducting SWNTs and vitamins B3 and C through density functional theory (DFT). At the same time, electronic transport calculations of these systems are performed using DFT combined with non-equilibrium Green's functions methods.

THEORETICAL DETAILS

The *ab initio* calculations of the VTB and VTC adsorptions on the SWNTs surfaces are based on the DFT calculations [4,5] using the SIESTA code [6]. We used (8,0) semiconducting SWNT with 64 atoms in the supercell. The double zeta basis set plus polarization functions (DZP) were employed and the exchange-correlation potential was adjusted by the local density approximation (LDA), according to the parameterization proposed by Perdew and Zunger [7]. To represent the charge density, a cutoff of 150 Ry for the grid integration in the real space was used. The interactions between ionic cores and valence electrons were described with norm-conserving pseudopotentials (Troullier-Martins) [8]. The structural optimizations were performed through the conjugate gradient algorithm [6] until the residual forces were smaller than 0.05 eV/Å.

To calculate the transport properties, we used a combination of the DFT with non-equilibrium Green's functions methods [9]. The system is initially separated in three parts, namely the left and right semi-infinite electrodes – which act as charge reservoirs – and the central scattering region containing a segment of the nanotube and the adsorbed molecule. The central quantity used to obtain the transport properties of a nanoscale device is the Green's function of the scattering region for a certain energy E

$$G = [E\,S - H - \Sigma_L - \Sigma_R\,]^{-1}, \tag{1}$$

where H and S are the Hamiltonian and overlap matrices for the central scattering region, respectively, and the effect of the semi-infinite electrodes is taken into consideration by self-energies Σ_L and Σ_R. The electronic transport properties can be subsequently calculated by a Landauer-type formula [9]. The densities of states can also be calculated using the Green's function formalism, thus avoiding spurious mini-gaps originating from periodic boundary condition calculations [10].

The binding energies (E_b) are calculated using the basis set superposition error (BSSE) [11]. This correction is done through the counterpoise method using "ghost" atoms within the following equation:

$$E_b = -[E_T(SWNT + X) - E_T(SWNT + X_{ghost}) - E_T(SWNT_{ghost} + X)] , \qquad (2)$$

where the X = VTC- or VTB- molecule radical and $E_T(SWNT + X)$ is the total energy of the SWCNT interacting with the radical. The "ghost" SWNT/X corresponds to additional basis wave functions centered at the position of the SWNT/X, but without any atomic potential.

DISCUSSION

The adsorptions of the VTB and VTC molecules on the (8,0) SWNT surface were done through the molecule functionalization where one hydrogen atom was removed of the amine and hydroxyl group, respectively. The Figures 1 (a) and (b) present the optimized structures for the studied systems.

The binding energies of the molecules radical and the SWNTs are evaluated using the equation (2). Covalent bonds are formed between the nanotubes and vitamins, with binding energies corresponding to 3.22 and 2.11 eV for VTB and VTC radicals, respectively. These values indicate that these systems may be manipulated in a rather stable way.

Using the Green's function formalism we proceeded to calculate the densities of states and projected densities of states of both vitamins B and C attached to the (8,0) SWNT. The results are shown in Figure 2. In both cases, a localized state can be seen in the middle of the gap of the (8,0) SWNT. In all cases, this localized state arises due to the interaction of the corresponding molecule with the SWNT and its character can be associated mostly with states from carbon atoms in the SWNT itself. In fact, we note that the height of the peak is much larger than the contribution due to the isolated molecule. It is possible also to observe that, apart from the small contribution to the gap states, the DOS height comes from the molecule and it is mostly located around -2.0 eV bellow the Fermi level.

The adsorptions of vitamins B and C onto the SWNT have a drastic effect on the electronic transport properties. The Figure 3 shows the zero bias transmission coefficients as a function of energy. In both cases, the effect of the molecule is stronger bellow the Fermi level instead of above it. In particular, even though there are no states associated to the molecule around the first conductance plateau, it is possible to observe strong scattering around that energy region. As discussed above, this is mainly due to localized states in the SWNT itself originating after the binding of the molecule radicals.

The signatures of the molecular states appear lower in energy, around 1.5 eV bellow the Fermi energy, in both cases. They appear in dip form in the conductance at energies comparable to peaks in the PDOS. These Fano-type resonances can be associated with the scattering events between the weakly coupled delocalized electrons in the nanotube and the localized molecular states in the molecule. We note that the dips are much sharper in the case of vitamin C, indicating that the binding between the molecules are weaker than in the case of vitamin B. This result is in accordance with our binding energy calculations and with the broadening of the energy levels, as seen in Figure 2.

181

CONCLUSIONS

In summary, first principles simulations and transport properties of the systems formed by the vitamins B3 and C molecules adsorbed on the (8,0) SWNTs are presented using density functional theory combined with non-equilibrium Green's functions methods. The adsorptions of these molecules are demonstrated to depend strongly on the functionalization of the adsorbed species. A chemisorption regime, via a strongly covalent bond, is observed for the functionalized vitamin molecule adsorbed on the nanotube. The transport properties of the SWNT change considerable when the vitamin radicals are adsorbed, showing significant reductions on the total transmissions, both at the valence and conduction bands, of the original SWNTs. It is also demonstrated that, in some cases, a sharp Fano-type resonance appears, indicating a weak coupling between the sharp states of the molecule and the block states belonging to the nanotube. The energetic and transport results presented in this work show that the SWNT, modified by adsorption of molecules on its surface, could result in promising vitamin carries, as well as chemical sensors or general drug delivery system.

ACKNOWLEDGMENTS

The authors acknowledge CENAPAD-SP for computer time and financial support from Brazilian agencies CNPq and FAPERGS (Grant 07/01129).

REFERENCES

1. S. Iijima, *Nature* **354**, 56 (1991).

2. S. K. Smart, A. I. Cassady, G. Q. Lu and D. J. Martin, *Carbon* **44**, 1034 (2006).

3. I. Zanella, S. B. Fagan, R. Mota and A. Fazzio, *Chem. Phys. Lett.* **439**, 348 (2007).

4. P. Hohenberg and W. Kohn, *Phys. Rev. B* **136**, 864 (1964).

5. W. Kohn and L. J. Sham, *Phys. Rev. A* **140**, 1133 (1965).

6. J. M. Soler, E. Artacho, J. D. Gale; A. García, J. Junquera, P. Ordejón and D. Sánchez-Portal, *Phys.: Condens. Matter.* **14**, 2745 (2002).

7. J. P. Perdew and A. Zunger, *Phys Rev. B* **23**, 5048 (1981).

8. N. Troullier and J. L. Martins, *Phys. Rev. B* **43**, 1993 (1991).

9. A. R. Rocha et al., Phys. Rev. B **73**, 085414 (2006); F. D. Novaes, A. J. R. da Silva, and A. Fazzio, Braz. J. Phys. **36**, 799 (2006).

10. A. R. Rocha, et al., Phys. Rev. B **77**, 153406 (2008).

11. S. F. Boys and F. Bernardi, *Mol. Phys.* **19**, 553 (1970).

Figure 1: Relaxed atomic structures for (a) VTB–radical and (b) VTC–radical configurations interacting with pristine SWNT.

Figure 2: Total density of states (DOS) and projected density of states for an (8,0) SWNT with a) vitamin B and b) vitamin C molecules attached to it.

Figure 3: Zero bias transmission coefficients as a function of energy for an (8,0) SWNT with a) vitamin B and b) vitamin C radicals molecule attached to it.

FIGURES

Figure 1

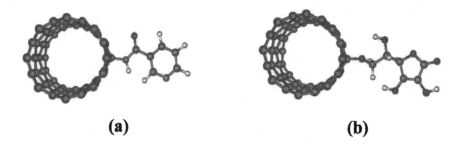

(a)   (b)

Figure 2

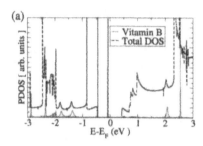

(a)

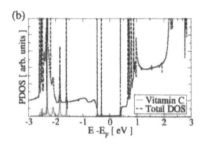

(b)

Figure 3:

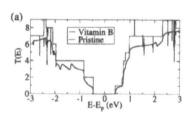

(a)

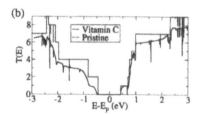

(b)

Mater. Res. Soc. Symp. Proc. Vol. 1142 © 2009 Materials Research Society   1142-JJ15-28

Addressable Carbon Nanotube Intra-Connects With Conductive Polymers

Seon Woo Lee[1] , Haim Grebel[1], Daniel Lopez[2] and A. Kornblit[2]
[1]Electronic Imaging Center at NJIT and the Electrical and Computer Engineering Department, New Jersey Institute of Technology (NJIT), Newark, NJ 07102
[2]New Jersey Nanotechnology Consortium (NJNC), Lucent Technologies Bell Labs, Murray Hill, NJ 07974

ABSTRACT

We have demonstrated gated structures made of individual single walled carbon nanotubes, which were electroplated with conductive polymers and measured their electrical and optical properties.

INTRODUCTION

Carbon nanotubes (CNT) and conducting polymer (CP) have studied extensively due to their remarkable electrical, optical, chemical, mechanical properties and diverse potential application as nano-scale devices.[1-4] Devices made of channeled CNT intra-connect have gained much interest owing to extremely narrow diameter. [5-7] In many past applications, conductive polymer based CNT devices were used for sensing, electronic and opto-electronic purposes. [8-12] However, only a few studies have been conducted for individually addressed channels owing to challenges involved in growing the carbon nanotubes precisely between pre-determined locations on the chip.[13] Here, we demonstrate electroplated gated devices, which were based on individual carbon nanotube intra-connects, spanning between two pre-determined electrode locations.

EXPERIMENT

Figure 1 shows the electrode configuration. Figure 1a shows a SEM image of the metal electrodes. The distance from tip to tip was 1 μm, although the mask allowed the electrode tips to co-align or laterally shifted (Figure 1a inset). Schematic of a field effect transistor (FET) made of individually addressable CNT intra-connect is shown in Figure 1b. The channel was grown by chemical vapor deposition[14,15] (CVD) technique and the two metal electrodes are used as source and drain electrodes, respectively. The silicon substrate was used as a back gate electrode. The silicon surface was oxidized to a thickness of 20 nm prior to the electrode deposition. Figure 1c shows the schematic of individually addressable CNT intra-connect electroplated with polycarbazole (PCZ). We have measured the source-drain characteristics for these CNT intra-connect before and after electrodeposition with PCZ. Photoconductance under white light illumination was assessed as well.

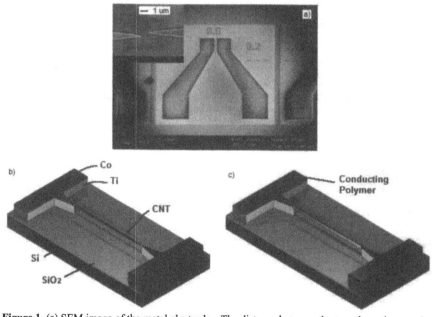

Figure 1. (a) SEM image of the metal electrodes. The distance between the two sharp tips was 1 μm. Individually addressed CNT intra-connect was grown directly from tip to tip by CVD. (b) Schematic diagram of our CNT FET. Co was deposited as a catalyst on top of Ti electrode to initiate CNT intra-connect growth. Si substrate was used as a back gate. (c) CNT intra-connect is electroplated by PCZ.

RESULTS AND DISCUSSION

An example of individually addressable CNT intra-connect is shown in Figure 2.

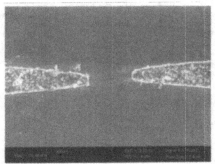

Figure 2. SEM image of CNT intra-connect.

Current- gate voltage I_{ds}-V_{gs} characteristics were measured for various V_{gs} from -10 V to +10 V. Negative differential resistance (NDR) was found in the I_{ds}-V_{gs} curves for gate bias in the region of $-3 > V_{gs} > -6$ V (Figure 3a). The NDR peak was shifted to the negative values as the source-drain voltage increased from V_{ds}=0 to 0.75 V. Otherwise, the intra-connects exhibited characteristics of an ordinary p-type channel.

The experiments were repeated under white light illumination (Figure 3b). The light increases the carrier density in the channel but not in the metal electrodes and allowed us to study the effective doping of the channel without affecting the work function of the SWCNT/metal contact. The overall channel conductance increased under the light irradiation. Under illumination, the devices became more stable, as well. While the origin of such behavior is unknown at this point, one may postulate that it may be attributed to screening owing to the similarity between the trends in Figure 3a and 3b.

Figure 3c and 3d are Ids-Vgs characteristics under dark and white light illumination conditions for the PCZ-coated CNT intra-connect. In general, there is a marked increase in the current for the electrodeposited CNT channels. On the other hand, the position of the NDR has not changed for these channels. This implies that the NDR effect ought to be attributed to the contact between the SWCNT and electrode. The NDR effect was somewhat masked for irradiated electroplated channels at low source-drain voltage. This points to the large conductivity difference between the carbon tube and the polymer, resulting in difference current distributions between these components under various voltages.

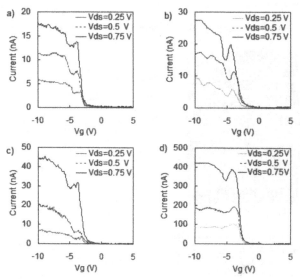

Figure 3. Electrical properties before and after electropolymerization. (a) I_{ds} vs. V_{gs} characteristics for CNT intra-connect. Negative differential resistance was observed for V_{gs} between -2 and -6. (b) I_{ds} vs. V_{gs} characteristics after white light illumination. (c) I_{ds} vs. V_{gs} characteristics after electroplating with PCZ under darkness. (d) I_{ds} vs. V_{gs} characteristics after electroplating with PCZ under white light illumination.

Figure 4 shows the channel response when the white light was turned on and off repeatedly. The uncoated channels exhibit almost an instantaneous response (Figure 4a). The electroplated channels exhibited a relatively long time constant during the OFF stage (Figure 4b). This is due to the charge hoping mechanism in these polymers.

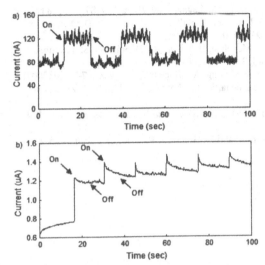

Figure 4. (a) Optical pulses are measured for CNT intra-connect. (b) For PCZ-coated CNT intra-connect.

CONCLUSIONS

In summary, we have demonstrated individual carbon nanotube channels which were electroplated by conductive polymers. The gated devised exhibited large negative differential resistance as well as a large photoconductance effects.

REFERENCES

1. M. F. Yu, O. Lourie, M. J. Dyer, K. Moloni, T. F. Kelly and R. S. Ruoff, *Science* **287**, 637-640 (2000).
2. J. W. G. Wildoer, L.C. Venema, A. G. Rinzler, R. E. Smalley and C. Dekker, *Nature* **391**, 59-62 (1998).
3. M. S. Dresselhaus, G. Dresselhaus and P. Avouris, "Carbon nanotubes: Synthesis, Structure, Properties and Applications", *Springer-Verlag*, Berlin (2001).
4. J. Li, Y. Lu and M. Meyyappan, *IEEE sensors journal* **6**, 1047 (2006).
5. J. Cao, Q. Wang, D. Wang and H. Dai, *Small* **1**, 138 (2005).
6. T.W. Tombler, C.W. Zhou, L. Alexseyev, J. Kong, H. J. Dai, L. Lei, C. S. Jayanthi, M. J. Tang and S. Y. Wu, *Nature* **405**, 769 (2000).
7. A. Javey, J. Guo, M. Paulsson, Q. Wang, D. Mann, M. Lundstrom and H. J. Dai, *Phys. Rev. Lett.* **92**, 106804 (2004).
8. X. Liu, J. Ly, S. Han, D. Zhang, A. Requicha, M. E. Thompson and C. Zhou, *Adv. Mater.* **17**, 2727 (2005).

9. Z. Xu, Y. Wu, B. Hu, I. N. Ivanov and D. B. Geohegan, *Appl. Phys. Lett.* **87**, 263118 (2005)
10. H. S. Woo, R. Czerw, S. Webster, D. L. Carroll, J. Ballato, A.E. Strevens, D. O'Brien, and W. J. Blau, *Appl. Phys. Lett.* **77**, 1393 (2000).
11. M. W. Rowell, M. A. Topinka, M. D. Mcgehee, H. J. Prall, G. Dennler, N. S. Sariciftci, L. Hu and G. Gruner, *Appl. Phy. Lett.* **88**, 233506 (2006)
12. M. C. Kum, K. A. Joshi, W. Chen, N. V. Myung, A. Mulchandani, *Talanta* **74**, 370 (2007).
13. X. Liu, J. Ly, S. Han, D. Zhang, A. Requicha, M. E. Thompson, and C. Zhou, *Adv. Mater.* **17**, 2727 (2005)
14. D. Katz D. Lopez, A. Kornblit and H. Grebel, *J. Nanotechnology and Nanostructures* **8**, 1-5 (2008).
15. D. Katz, S-W. Lee, D. Lopez, A. Kornblit and H. Grebel, *J. Vac Sys. and Tech. B* **25**, 1191 (2007).

Mater. Res. Soc. Symp. Proc. Vol. 1142 © 2009 Materials Research Society   1142-JJ15-41

A New Thermionic Cathode Based on Carbon Nanotubes With a Thin Layer of Low Work Function Barium Strontium Oxide Surface Coating

Feng Jin, Yan Liu, Scott A Little and Christopher M Day
Department of Physics and Astronomy, Ball State University, Muncie, IN 47306

ABSTRACT

We have created a thermionic cathode structure that consists of a thin tungsten ribbon; carbon nanotubes (CNTs) on the ribbon surface; and a thin layer of low work function barium strontium oxide coating on the CNTs. This oxide coated CNT cathode was designed to combine the benefits from the high field enhancement factor from CNTs and the low work function from the emissive oxide coating. The field emission and thermionic emission properties of the cathode have been characterized. A field enhancement factor of 266 and a work function of 1.9 eV were obtained. At 1221 K, a thermionic emission current density of 1.22A/cm^2 in an electric field of 1.1 V/μm was obtained, which is four orders of magnitude greater than the emission current density from the uncoated CNT cathode at the same temperature. The high emission current density at such a modest temperature is among the best ever reported for an oxide cathode.

INTRODUCTION

Carbon nanotubes (CNTs) are natural field emitters; their unique geometry and high aspect ratio give rise to a high field enhancement factor . The field emission properties of CNTs have been extensively studied in recent years. [1-4] However, there are few reports on the thermionic emission properties of CNTs in the literature. The benefit of the large field enhancement factor introduced by CNTs has not been exploited for thermionic emission and thermionic cathode applications. [5-7]

The governing equation for thermionic emission is the Richardson-Dushman equation:

$$J_s = 120T^2 e^{-11605\varphi/T} e^{4.4\sqrt{E}/T} \qquad (1)$$

The last exponential term in the equation represents the field effect in thermionic emission and is referred to as the Schottky Effect. The Schottky Effect is usually not very large and has been largely overlooked. However, with the aid of modern noanotechnology, it is possible to create a much larger Schottky effect by introducing a large field enhancement factor, which could potentially lead to a dramatic increase in thermionic emission.

In this paper, we report a thermionic cathode structure based on CNTs, and its electron emission properties. The cathode consisted of three major components: the metal substrate, which was a thin tungsten ribbon that could be heated as a filament by flowing through an electric current; CNTs on the surface of the tungsten ribbon that provided a large field enhancement factor; and a thin layer of low work function oxide (BaO/SrO) materials. The basic idea of this oxide coated CNT cathode structure was to combine the benefits of the large field enhancement factor introduced by CNTs and the low work function from the oxide coating to improve the overall thermionic electron emission.

EXPERIMENTAL DETAILS

CNTs were grown on the tungsten ribbon in a region approximately 0.012 cm^2 in area using plasma enhanced chemical vapor deposition (PECVD). A thin film of nickel, typically about 100 nm, was first sputter deposited on the surface of the tungsten ribbon. The ribbon was then transferred to a PECVD system and etched in NH$_3$ plasma at 600 °C for several minutes. This was followed immediately by flowing C$_2$H$_2$ to start the CNT growth. The ratio of C$_2$H$_2$ to NH$_3$ was 1:2 and the flow rate was kept at 240 sccm (standard cubic centimeters per minute). For both plasma etching and CNTs growth, the power of the DC plasma was 30 W, the gas pressure in PECVD was maintained at approximately 6 Torr by adjusting the opening of the gate valve of the PECVD system, and substrate temperature was maintained at 600 °C. The length of the CNTs was controlled by the growth time, the growth time used for the samples presented here were around 30 minutes. The diameter and spacing of the CNTs were controlled by the thickness of the initial nickel thin film and the NH$_3$ plasma etching that preceded CNT growth. Samples of CNTs with various surface morphologies were produced by controlling these growth parameters. The CNTs were examined with scanning electron microscopy (SEM) and transmission electron microscopy (TEM).

Barium strontium oxide (BaO/SrO), a common emissive materials system for oxide cathodes, was used for the low work function oxide coating materials, and was deposited onto the surface of CNTs using the magnetron sputter deposition technique. The work function of BaO/SrO coatings typically range from 1 to 2 eV.

Field emission and thermionic emission from the CNT and the oxide coated CNT cathodes were characterized in an ultrahigh vacuum (< 5x10^{-8} Torr) chamber in a simple diode configuration. The test emitters were mounted on one side and used as the cathodes. On the opposite side a copper plate was used as the anode. The spacing between the cathode and anode was controlled and adjusted by a micrometer. A high voltage source-measurement unit was used to generate a high voltage across the cathode and anode. I-V (or I-E) curves for emitters were obtained by measuring the electron emissions from the cathode at different anode voltages. For the thermionic emission measurement, another source-measurement unit was used to resistively heat the cathode to the desired temperatures by varying the current flow through the tungsten ribbon. The temperature of the cathode was determined using an optical pyrometer. At each temperature, the current from electron emission was measured for different values of anode voltage or field strength.

DISCUSSIONS

Figures 1(a), (b) and (c) are TEM images of individual CNTs, nickel coated CNTs and oxide coated CNTs respectively. The metal and oxide surface coatings can be clearly seen from Fig. 1(b) and (c). The thickness of the oxide coating was approximately 100 nm thick. The oxide coating was fairly uniform and covered the whole surface of the CNTs. The overall shape of the coated CNTs was similar to that of the uncoated CNTs, while their diameters were increased due to the additional surface coatings. The formation of the oxide

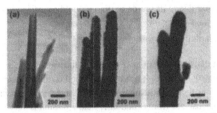

Fig. 1. TEM images of uncoated CNTs (a), 50 nm nickel coated CNTs (b), and 100 nm barium strontium oxide coated CNTs (c).

coating on the CNT surfaces was confirmed with chemical analysis using energy-dispersive spectrometry (EDS). Fig. 2(a), (b) and (c) show the SEM images of the uncoated CNTs, nickel coated CNTs and oxide coated CNTs respectively. The CNT samples used here were from the same batch as sample A in the previous section. The diameters of the oxide coated and nickel coated CNTs appeared thicker than the uncoated ones in the SEM images, while the overall surface morphology and the alignment of the coated CNTs remained largely unchanged by the deposition of the surface coatings. These TEM and SEM images clearly demonstrated the ability of the magnetron sputtering technique to deposit uniform thin films on extremely uneven surfaces like CNTs.

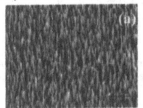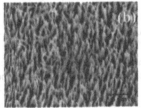

Fig. 2. SEM images of uncoated CNTs (a), and 100 nm barium strontium oxide coated CNTs (c). (Tilted 45° view).

The field emission measurement results are shown in Fig. 3. Fig. 3(a) contains the I-E plots showing the dependencies of field emission current on electric field for the CNT emitter and the oxide coated CNT emitter.

The benefit of the low work function oxide coating can be clearly seen in Fig. 3(a). The field emission of the oxide coated CNT emitter was approximately 23.6 μA at 4.4V/μm, more than twice that of the uncoated CNT emitter. Fig. 3(b) shows the Fowler-Nordheim (FN) plots for the two emitters. The β factors for these two emitters were calculated from the slopes of the F-N plots. The work function values used for the calculation were the measured values of 1.9 eV and 4.5 eV for the oxide coated CNT emitter and the CNT emitter respectively. The work function measurement is described in the later section. A fairly large β of 1484 was obtained for the CNT emitter, and a smaller yet still significant β of 467 was obtained for the oxide coated CNT emitter. This reduction of the β factor was a result of the

increase of tube diameter due to the oxide coating. Field emission is described by the
Fowler-Nordheim equation:

$$J = \eta a \frac{(\beta F)^2}{\phi} \exp(-b \frac{\phi^{3/2}}{\beta F})$$ (2)

where J is the current density in A/cm^2; ϕ is the work function in eV; $a = 1.54 \times 10^{-6}$ A eV V^{-2} and $b = 6.83 \times 10^7$ (eV)$^{-3/2}$ V cm^{-1}; η is a factor that describes the geometrical efficiency of electron emission; β is the field enhancement factor; and F is the macroscopic electrical field in V/cm. As one can see from equation 1, field emission is very sensitive to the β factor. Given this strong dependence, a decrease in β of this magnitude would normally have resulted in a significant reduction of field emission current. However, the work function is also a key factor that determines the field emission from an emitter. In this particular example, the lowering of the work function from 4.5 eV to 1.9 eV overcame the effect of the reduction of β from 1484 to 467, resulting in an overall increase of field emission.

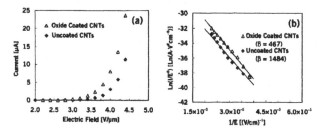

Fig. 3. Field emission current as a function of electric field, comparing emission from the uncoated CNTs to oxide coated CNTs (a), and the corresponding Fowler-Nordheim plots (b).

As mentioned earlier, the governing equation for thermionic emission is the Richardson-Dushman equation:

$$J_s = 120 T^2 e^{-11605\varphi/T} e^{4.4\sqrt{E}/T}$$ (3)

where the term J_s indicates that this equation refers to the saturation current; T is the temperature in Kelvin; ϕ is the work function in eV; E is the external electric field in V/cm; and A is Richardson's constant, with a value of 120 A/K^{2}cm^2. The second exponential term in the equation represents the field effect in thermionic emission and is referred to as the Schottky effect. The first exponential and the preceding constants are called the zero field emission: The determining factors for the thermionic emission are temperature and work function. The Schottky effect is usually not very large. However, a much larger field effect is possible by introducing a large field enhancement factor, which could potentially lead to a dramatic increase in thermionic emission, and this is where CNTs came into the play. Fig. 4(a) and (c) show the thermionic emission current density dependencies on electric field at various temperatures for the CNT emitter and the oxide coated CNT emitter respectively. Fig. 4(b) and (d) contain the corresponding Schottky plots for the CNT and oxide coated CNT emitters. Compared to the CNT emitter, the oxide coated CNT emitter produced thermionic emission at a much lower temperature. As predicted by the Schottky law, the emissions entered the saturation region at high electric fields and became linearly dependent on the square root of electric field for both the CNT and oxide coated CNT emitters. The zero field emissions at different temperatures were extrapolated and determined from the Schottky plots, which were then used to generate the Richardson plots. The Richardson plots for the CNT and oxide coated CNT emitters in Fig. 4(e) fit straight lines well, and the work functions of the samples were calculated from the slopes of the lines. The calculated work function value was 1.9 eV for the oxide coated CNT emitter and 4.5 eV for the CNT emitter, confirming a significant reduction of the work function due to the BaO/SrO thin film surface coating. The benefit of the low work function oxide coating is clearly demonstrated in the thermionic emission current density plots in Fig. 4(a) and (b). The thermionic emission from the oxide coated

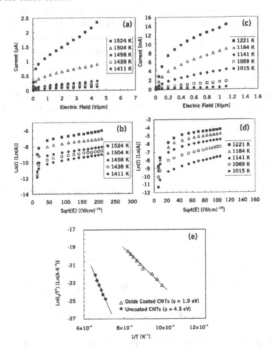

Fig. 4. The field dependence of thermionic emission current at various temperatures for the CNT emitter (a), the corresponding Schottky plots for the CNT emitter (b), the field dependence of thermionic emission current at various temperatures for the oxide coated CNT emitter (c), the corresponding Schottky plots for the oxide coated CNT emitter (d), and the Richardson plots for both the oxide coated CNT and uncoated CNT emitters (e). The work functions calculated from the Richardson plots are 4.5 eV for the CNT emitter and 1.9 eV for the oxide coated emitter.

CNT emitter was more than four orders of magnitude higher than the thermionic emission from the uncoated CNT emitter. This jump in emission current and emission current density was the result of reducing the work function from 4.5 eV to 1.9 eV. At 1221 K and an electric field of 1.1 V/μm, a 14.6 mA emission current was obtained from the oxide coated emitter in constant DC operation mode. The area of the emission surface was 0.012 cm^2. An emission current density of 1.22 A/cm^2 was calculated based on the emission current and the emitting area.

CONCLUSIONS

In summary, we have created a highly efficient field enhanced themionic cathode based on oxide coated CNTs. The cathode structure is such that it takes advantage of both the large field effect introduced by CNTs and the low work function of oxide surface coating. The resulting oxide coated CNTs cathode had a work function of 1.9 eV and a field enhancement factor of 467, which led to a significant improvement in both field and thermionic emission. Compared to uncoated CNTs, the field emission was increased by a factor of two, while the thermionic emission increased by more than four orders of magnitude. At 4.4 V/μm, a field emission current of 23.6 μA was obtained from an emitting surface of 0.012 cm^2. Similarly, at of 1221 K and an electric field of 1.1 V/μm, a 14.6 mA emission current and 1.22A/cm^2 emission current density were obtained from the oxide coated emitter in CW operation mode. This is one of the highest emission current densities from oxide cathodes ever reported at such a modest temperature.

ACKNOWLEDGEMENT

This work was supported by the Department of Energy under contract No. DE-FC26-04NT42329.

REFERENCES

1. De Jonge N, Bonard JM. Carbon nanotube electron sources and application. Phil Trans R Soc Lond A 2004; 362(1823):2239-2266.
2. De Jonge N, Brightness of carbon nanotube electron sources. J Appl Phys 2004; 95(2):673-681
3. De Heer WA, Chatelain A, Ugarte D. A carbon nanotube field-emission electron source. Science 1995; 270:1179-1180
4. Cheng Y, Zhou O. Electron field emission from carbon nanotubes. C R Physique 2003; 4:1021–1033
5. Cox DC, Forrest RD, Smith PR, Silva SRP. Thermionic emission from defective carbon nanotubes. Appl Phys Lett 2004; 85(11):2065-2067
6. Shiraishi M, Ata M. Work function of carbon nanotubes. Carbon 2001; 39(12):1913-1917
7. Jin F, Liu Y, Day CM. Thermionic emission from carbon nanotubes with a thin layer of low work function barium strontium oxide surface coating. App Phys Lett 2006; (88):163116 1-3.

Mater. Res. Soc. Symp. Proc. Vol. 1142 © 2009 Materials Research Society   1142-JJ15-42

Characterization of Pt Nanocontacts to ZnO Nanowires Using Focused-Ion-Beam Deposition

Pei-Hsin Chang, Kun-Tong Tsai, Cheng-Ying Chen and Jr-Hau He
Institute of Photonics and Optoelectronics, and Department of Electrical Engineering, National Taiwan University, Taipei, 106 Taiwan (ROC)

ABSTRACT

Understanding the transport across the contact between a metal electrode and ZnO is a key issue to fabricating high performance ZnO nanowire-based nanodevices. In this study, we have characterized the contact between the focused-ion-beam-microscopy-deposited Pt and ZnO nanowires. The dominant transport mechanism of the contact is the thermionic field emission (TFE) process. It is found that the presence of Ga plays an important role to tune the thermionic emission into TFE transport. The discovered phenomena and underlying mechanisms are not only of broad scientific interests but also of great technological significance, since understanding the transport of semiconductor nanostructures paves the way to fabricate high performance of nanowire-based devices.

INTRODUCTION

Due to the ultrahigh surface-to-volume ratio characteristics of nanowires, significant progress has been made in the application of ZnO nanostructures [1] for the fabrication of various electronic, optoelectronic, and sensor devices which include piezoelectric nanogenerators [2], chemical sensors [3], spin functional devices[4], nanolasers [5], and dye-sensitized solar cells[6], photodetectors [7,8], piezoelectric gated diodes [9]. Since the performance of nanodevices critically depends on the quality of the Ohmic contacts with electrodes, the development of low resistance contacts is vitally important for fully utilizing the advantages offered by ZnO [10]. A high contact resistance can greatly degrade the performance of the device, resulting in gigantic reduction or elimination of the effects or benefits provided by nanomaterials. A low resistance Ohmic contact can be achieved either through surface preparation to reduce the metal semiconductor barrier height, increasing carrier tunneling probability, or by increasing the effective carrier concentration of the surface [11].

Focused ion beam (FIB)-based nanofabrication, which involves milling, implantation, ion-induced deposition, and ion-assisted etching of materials, has become an increasingly popular tool for the fabrication of various types of nanostructures for different applications [12,13]. To form low-resistance contacts on ZnO materials at room temperature, FIB, a rapid and flexible method, has been widely utilized to deposit Pt metal as the contacts on n-type ZnO nanowire-based nanodevices [14-17]. However, there exists a Schottky barrier between the pure Pt and ZnO since the electron affinity of ZnO is 4.5 eV and the work function of Pt metal is 5.65~6.1 eV [2,18]. The deposition in FIB inevitably involves Ga ions. Therefore, understanding the transport across the contact between FIB-deposited Pt (FIB-Pt) and ZnO is the key issue to fabricate high performance of ZnO nanowire-based nanodevices.

In this work, the FIB-Pt contact with ZnO nanowires is studied in details for the first time. The specific contact resistance has been determined as low as 9.4×10^{-6} Ωcm^2 using the four-terminal Kelvin measurement. To further clarify the transport behavior across the Schottky barrier between the ZnO nanowire and FIB-Pt, the measurement of temperature-dependent contact resistance has been performed. The transport mechanism of the FIB-Pt contacts on the ZnO nanowires is suggested to be dominated by the thermionic field emission (TFE). Structure analysis using transmission electron microscopy (TEM) and related analytical tools proves that the presence of Ga plays an important role to tune the thermionic emission (TE) into TFE transport.

EXPERIMENT

Details of the vapor-liquid-solid growth technique employed for the fabrication of the ZnO NBs has been reported elsewhere [19]. The synthesized ZnO nanowires are transferred from the Si substrate to SiO$_2$ substrate that was prepatterned with Au/Ti electrodes. Gallium ion beam induced deposition in a FEI Dual-Beam NOVA 600 FIB instrument is employed with a trimethylcyclopentadienyl-platinum $(CH_3)_3Pt(C_pCH_3)$ injector to selectively deposit Pt metal to connect Au/Ti electrodes with ZnO nanowire. The Ga$^+$ ion beams is accelerated to 30 kV, injecting perpendicular to the ZnO nanowires, and the income flux was set at 50 pA during the deposition. The dimensions of Pt deposition areas were confined ~300 nm in width and 300 nm in height.

The contact resistance between a FIB-Pt electrode and a ZnO nanowire is measured by a four-probe Kelvin measurement using Keithley 236 Sourcemeter. A typical SEM image of the test structure of the four-probe measurement is shown in Figure 1. By this method, the high impedance of two inner terminals would minimize the current flow through two inner terminals. Thus, since there is no potential drop across the contact resistance of two inner terminals, only the resistance of the semiconductor can be measured. Representative two-point and four-point I-V curves at the vicinity of V = 0 are shown in Figure 2. In addition to the nearly Ohmic behavior around V = 0 for the contacts in both curves, one can see that the current levels of the two-point and four-point measurements are at the same order, suggesting low specific contact resistance produced by the FIB-Pt. The detailed four-probe Kelvin measurement is described as follows. As shown in Figure 1(b), two outer terminals are used to source a current I_{14} through the ZnO nanowire, and two inner terminals are used to sense the potential drop V_{23} between the two inner points on the nanowire. Using the measurement of V_{23} and I_{14}, the resistance of the segment between the two inner points of the ZnO nanowire, $R_{NW} = \frac{V_{23}}{I_{14}}$, can be obtained. Using two-probe method, the total resistance (R_T) between the two inner terminals can be obtained. Thus the contact resistance (R_c) can be determined using

$$R_T = 2R_c + R_{NW} \qquad (1)$$

Finally, the R_c can be obtained. The specific contact resistance (ρ_c) is defined to be $\rho_c = R_c \times A$, where A is the active area of contact. We assume that the 75 % circumference of the nanowire of radius r is in direct contact with FIB-Pt metal. Therefore, A is estimated to be $0.75 \times (2\pi r w) = 1.5\pi r w$, where w is the width of the FIB-Pt metal line. Note that an estimation of the effective contact area is a rough approximation according to TEM observation. The specific contact resistance measured by four-probe method is 9.4×10^{-6} Ωcm^2 at room temperature.

Figure 1. (a) SEM image of a ZnO nanowire with FIB-Pt contacts. (b) Schematic of measurement pattern of the ZnO nanowires for four-terminal Kelvin measurement. I_1 and I_4 terminals source a current, and the potential difference is measured across V_2 and V_3 terminals.

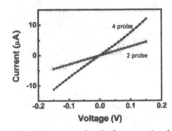

Figure 2. Two-point and four-point I-V curves for ZnO nanowire for low driving field, which is needed for estimating the local contact resistance. The two-point curve is taken by varying the voltage and measuring current. The four-point curve is taken by varying the current while measuring voltage, and inverting.

DISCUSSION

To further characterize the good contact between the FIB-Pt with ZnO nanowire, structural analysis in contact region has been carried out using a JEOL JEM 2100F scanning transmission electron microscopy (STEM)/TEM. The TEM sample was prepared by slicing the contact region using FIB. Figure 3 shows a typical TEM image of the FIB-Pt contact to ZnO nanowire as viewing along the direction of the nanowire. The contrast of FIB-Pt is due to the different local thickness of Pt [20]. As show in Figure 3(a), due to a difference in milling rate, the ZnO nanowire is rather thick [20]. Figure 3(b) confirms that ZnO is of the wurtzite structure. The diffused ring-shaped SAED pattern (Figure 3(c)) indicates that FIB-deposited metal is composed of nanocrystallite. Figure 3(d) is a high-resolution TEM (HRTEM) image from nanowire, which indicates that the nanowire is single crystalline. The HRTEM image of FIB-Pt region (Figure 3(e)) confirms that FIB-Pt is consisted of nanocrystallites and disordered materials.

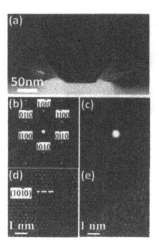

Figure 3. (a) Cross-sectional TEM image of ZnO nanowire, (b) the corresponding diffraction patterns of ZnO nanowire, (c) the corresponding diffraction patterns of FIB-Pt, (d) HRTEM image of ZnO nanowire, and (c) HRTEM image of FIB-Pt.

Energy dispersive X-ray spectroscopy (EDS) mapping is employed to investigate the distribution of the chemical compositions. The bright-field STEM image in Figure 4(a) shows that the nanowire is covered by the FIB-Pt on the SiO_2/Si substrate. ZnO is composed of Zn and O, as shown in Figures 4(b) and 4(c). Note that quantitative analysis of EDS is not accurate for light elements, such as C and O due to low X-ray fluorescence yields. The EDS mapping of Pt shows that the FIB-deposited contact contains Pt (Figure 4 (d)). Figure 4(e) shows that a certain amount of Ga exhibits in the FIB-Pt contact and junction of FIB-Pt/ZnO. There are two possible Ga sources accounting for this distribution. First, Ga-ion scanning is inevasible to be utilized for imaging microscope prior to Pt deposition in FIB. The surfaces of the ZnO would be modified under the Ga^+ ion irradiation. The dose value of Ga^+ ion beam is estimated by the scanning frame (42µm×48µm) of the Ga^+ ion beam at 512×442 pixels at 50 pA current. The duration time at every pixel is 3 µs. The substantial dose on the area of a frame is:

$$\left[\left(512 \times 442 \times 50\,pA \times 3\,\mu s \right) \Big/ 1.6 \times 10^{-19} \right] \Big/ 42\,\mu m \times 48\,\mu m = 0.9 \times 10^{13} \ (1/cm^2)$$

Secondly, due to the characteristics of FIB-induced deposition mechanism (i.e. Ga ion beam induced deposition), it is impossible to avoid the influence of the Ga ion, which was inevitably incorporated into the FIB-Pt contact during the Pt deposition process. Therefore, the data suggest the presence of Ga in the Pt contact, forming point defects in Pt. The Ga doping in ZnO nanowires might be responsible for the unusual low contact resistance of FIB-Pt to ZnO nanowires. The electrical characterization will be performed to confirm this suggestion further.

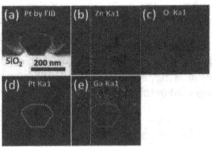

Figure 4. (a) Cross-sectional STEM image of ZnO nanowire, and the corresponding EDS mapping of (b) Zn, (c) O, (d) Pt, and (e) Ga.

The electron affinity of ZnO is 4.5 eV, and the work function of Pt metal is 5.65~6.1 eV [21], thus there exists a Schottky barrier at the Pt and ZnO interface [2,18]. The current flowed through in a Schottky barrier junction is dominated by the transport of the carriers from semiconducting ZnO to Pt metal. The transport can occur by different possible mechanisms, such as (1) thermionic emission (TE) over the barrier, (2) thermionic field emission (TFE) assisted by interface states existing in the metal-semiconductor junction, and (3) field emission (FE). While FE is a pure tunneling process, TFE is tunneling of thermally excited carriers which meet a thinner barrier than FE. The specific contact resistance is a function of the barrier height, carrier concentration (in TFE and FE mechanisms), and the temperature (in TE and TFE mechanisms) [22]. Since FIB-Pt metal forms low-resistance contacts with the ZnO nanowires, we deduce that the transport mechanism for contact of the FIB-Pt to ZnO nanowires could be the TFE or FE mechanism instead of TE (Schottky contact). For further investigating the transport mechanism between the FIB-Pt and ZnO interface, the measurement of the temperature-dependent specific contact resistance were performed in the Lakshore TTP4 cryogenic probe station at 5×10^{-6} torr., as shown in Figure 5. It is found that the specific contact resistance of FIB-Pt to ZnO nanowire decreased with increasing temperature. Accordingly, we conclude that the electrical transport of the Pt contacts on ZnO nanowire is the TFE mechanism since FE mechanism is independent of temperature. As the TFE mechanism dominates the transport of contact, the specific contact resistance when V → 0 is given by [22]

$$\rho_c = \frac{k\sqrt{E_{00}}\cosh\left(E_{00}/kT\right)\coth\left(E_{00}/kT\right)}{A^{**}Tq\sqrt{\pi q\left(\phi_{Bn}-\phi_n\right)}}\exp\left[\frac{q\left(\phi_{Bn}-\phi_n\right)}{E_{00}\coth\left(E_{00}/kT\right)}+\frac{q\phi_n}{kT}\right]$$

$$E_{00} = \frac{qh}{4\pi}\sqrt{\frac{N_d}{\varepsilon_0\varepsilon_r m^*}} \qquad (2)$$

TFE occurs at the energy above the conduction band, where ϕ_n is negative for degenerate semiconductor, E_{00} is the Padovani–Stratton parameter that depends on the semiconductor type and the carrier concentration at the interface (N_d), m^* is the effective mass of the ZnO, ε_0 is the vacuum dielectric constant, ε_r is the relative dielectric constant of the ZnO semiconductor.

According to equations (2), the specific contact resistance is a function of barrier height, carrier concentration, and temperature. We assume the barrier height of the contact, and carrier concentration at the interface are the constants. Thus the specific contact resistance is varied with temperature. The fitting curve according to equation (2) is consistent with the data we measured, as shown in Figure 5. By the fitting constants of the curve, E_{00} is calculated to be 0.35 meV. Substituting the constants of m^*, ε_0, and ε into equation (2), the carrier concentration at the interface can is estimated to be ~1.0×10^{15} (1/cm³).

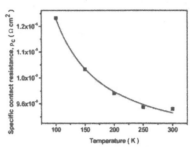

Figure 5. Temperature dependence of specific contact resistance using four-probe Kelvin measurement. The squares are the data acquired from measurement. The solid line is the fitting curve according to equation (2).

The origin of estimated carrier concentration in the ZnO nanowires is attributed to the addition of Ga during Ga ion imaging. The distributions of Ga atoms are calculated for ZnO nanowire by the TRIM code, which predicts a mean projected range of the Ga⁺ ions of ~15 nm. Thus most implanted Ga ions should be contained at the surface of the nanowires despite a non-uniform distribution. To compare the implanted dose with the volume density of implanted Ga⁺ ions, we assume a penetration depth of 200 nm is equal to the diameter of the nanowire as a rough estimate, which corresponds to an implanted Ga concentration of 4.5×10^{17} cm⁻³ for a dose of 0.9×10^{13} cm⁻². Ga doping concentration resulted from Ga ion imaging is two orders higher than the calculated carrier concentration (1.0×10^{15} cm⁻³) since it has been investigated that not all ions are electrically active as expected for as-implanted samples without any heat treatments [24]. It concludes that the TFE transport is contributed from the addition of Ga during Ga ion imaging. Our characterizations provide the explanation as to why FIB-Pt exhibits abnormal low-resistance contacts to ZnO nanowires. The presence of Ga in the ZnO makes the corresponding width of the barrier decreases and the thermally excited carriers tunnel through rather than over the potential barrier. Moreover, it has been known that metal screening lengths are negligible (Angstrom scale) compared to semiconductors (tens to hundreds of nanometers). The barrier width is modulated by doping within the semiconductor instead of metal [25]. The varied barrier width in the Pt due to the addition of Ga can be neglected. In addition, there will be interest in how the transport varies as the concentration of Ga in ZnO nanowires is varied to form lower contact resistance to ZnO nanowire-based devices. The effect of Ga doping densities on the contact resistance of Pt metals to ZnO nanowires and other nanomaterials is currently under investigation.

CONCLUSIONS

In conclusion, the specific contact resistance of the FIB Pt contacts on the ZnO nanowires has been determined as low as 9.4×10^{-6} Ωcm^2 using the four-terminal Kelvin measurement. The dominant transport mechanism of the FIB-Pt contact to the ZnO nanowire is suggested to be thermionic field emission, resulted from the presence of Ga in the ZnO. The carrier concentration in the ZnO can be estimated to be $\sim1\times10^{15}$ ($1/cm^3$). The discovered phenomena and underlying mechanisms are not only of broad scientific interests but also of great technological significance, because understanding the transport of semiconductor nanostructures paves the way to fabricate high performance of nanowire-based devices.

ACKNOWLEDGMENTS

The research was supported by the National Science Council Grant No. NSC 96-2112-M-002-038-MY3 and NSC 96-2622-M-002-002-CC3, and Aim for Top University Project from the Ministry of Education.

REFERENCES

1. Z. L. Wang, *J. Nanosci. Nanotech.* **8**, 27 (2008).
2. Z. L. Wang, J. H. Song, *Science* **312**, 242 (2006).
3. M. Law, H. Kind, B. Messer, F. Kim, P. D.Yang, *Angew. Chem. Int. Edit.* **41**, 2405 (2002).
4. S. P. Anthony, J. I. Lee, J. K. Kim, *Appl. Phys. Lett.* **90**, 103107 (2007).
5. M. H. Huang, S. Mao, H. Feick, H. Q.Yan, Y. Y. Wu, H. Kind, E. Weber, R. Russo, P. D. Yang *Science* **292**, 1897(2001).
6. M. Law, L. E. Greene, J. C. Johnson, R. Saykally, P. Yang *Nat. Mater.* **4**, 455(2005).
7. Chang C. Y., F. C. Tsao, C. J. Pan, G. C. Chi, H. T. Wang, J. J. Chen, F. Ren, D. P. Norton, S. J. Pearton, K. H. Chen, L. C. Chen, *Appl. Phys. Lett.* **88**, 173503 (2006).
8. J. H. He, S. T. Ho, T. B. Wu, L. J. Chen, Z. L. Wang, *Chem. Phys. Lett.* **435**, 119 (2007).
9. J. H. He, C. L. Hsin, J. Liu, L. J. Chen, Z. L. Wang *Adv. Mater.* **19**, 781 (2007).
10. O. Kordina, J. P. Bergman, A. Henry, E. Janzen, S. Savage, J. Andre, L. P. Ramberg, U. Lindefelt, W. Hermansson, K. Bergman, *Appl. Phys. Lett.* **67**, 1561 (1995).
11. J. J. Chen, S. Jang, T. J. Anderson, F. Ren, Y. Li, H. S. Kim, B. P. Gila, D. P. Norton, S. J. Pearton, *Appl. Phys. Lett.* **88**, 3 (2006).
12. C. A. Volkert, A. M. Minor, *MRS Bull.* **32**, 389 (2007).
13. A. A. Tseng, *Small* **1**, 924 (2005).
14. J. H. He, Y. H. Lin, M. E. McConney, V. VTsukruk., Z. L. Wang, G. Bao *J. Appl. Phys.* **102**, 084303 (2007).
15. C. Xu, S. Youkey, J. Wu, J. Jiao, *J. Phys. Chem. C* **111**, 12490 (2007).
16. R. Zhu, D. Q. Wang, S. Q. Xiang, Z. Y. Zhou, X. Y. Ye, *Nanotechnology* **19**, 5 (2008).
17. L. Liao, H. B. Lu, J. C. Li, C. Liu, D. J. Fu, Y. L. Liu, *Appl. Phys. Lett.* **91**, 173110 (2007).
18. W. I. Park, G. C. Yi, J. W. Kim, S. M. Park, *Appl. Phys. Lett.* **82**, 4358 (2003).
19. J. H. He, J. H. Hsu, C. W. Wang, H. N. Lin, L. J. Chen, Z. L. Wang, *J. Phys. Chem. B* **110**, 50 (2006).
20. D. Tham, C. Y. Nam, J. E.Fischer, *Adv. Mater.***18**, 290 (2006).
21. C. Y. Nam, D. Tham, J. E. Fischer, *Nano Lett.* **5**, 2029 (2005).

22. J. Liu, P. Fei, J. Song, X.D. Wang, C. Lao, R.Tummala, Z. L. Wang, *Nano Lett.* **8**, 328 (2008).
23. S. M. Sze, "*Physics of semiconductor devices,*" Wiley: New York, 1981.
24. H. Ryssel, I. Ruge, "*Ion Implantation,*" Wiley & Son: Chichester 1986.
25. X. Wu, E. S.Yang, *IEEE Electron Device Lett.* **11**, 315 (1990).

Mechanical Properties

Mater. Res. Soc. Symp. Proc. Vol. 1142 © 2009 Materials Research Society 1142-JJ16-04

The Mechanical Response of Aligned Carbon Nanotube Mats via Transmitted Laser Intensity Measurements

Christian P. Deck[1], Chinung Ni[1], Kenneth S. Vecchio[2], and Prabhakar Bandaru[1]
[1]Materials Science and Engineering, U. of California, San Diego, La Jolla, Ca, 92093-0418, USA
[2]Dept. of Nanoengineering, U. of California, San Diego, La Jolla, Ca, 92093-0448, USA

ABSTRACT

Carbon nanotubes are one of the more widely studied nanostructures today, with ongoing attempts to exploit their small size, large aspect ratio, and combination of mechanical, optical, and electronic properties. In this work, a chamber was designed to pass a laser through a mat of aligned carbon nanotubes and monitor the variation in transmitted light intensity in response to different mechanical deformations. This approach specifically takes advantage of the scale and mechanical and optical properties of carbon nanotubes, particularly their high elastic limit and anisotropic light absorption.

In this study, we measure the flexural rigidity (the product EI) of carbon nanotube arrays. Vertically aligned nanotubes were grown in periodic arrays, and fluid flow was applied normal to the nanotube axis. These nanotubes deflect due to shear caused by fluid drag, and this deflection is monitored experimentally with high accuracy by measuring a decrease in transmitted light intensity as a function of increasing fluid velocity and density. This response was also simulated, using a model based on the Stokes-Oseen equations with a correction for the small length scales associated with nanotube mats. The experimental deflection data and the estimated force on the tubes from simulations are used for determining the flexural rigidity of CNTs, to be of the order of 10^{-15} Nm^2. Using this method, we also demonstrate a carbon nanotube-based fluid flow and shear force sensor that offers fast response, repeatability, and that can measure forces in very close proximity to surfaces (such as in boundary layer flow).

INTRODUCTION

The properties of carbon nanotubes (CNTs)[1] have been extensively studied. In particular, the exceptional mechanical properties of nanotubes, which include high strength,[2] stiffness,[3] and elastic limit,[4] are of great interest. These properties have been investigated using a number of theoretical and experimental methods, including molecular dynamics simulations,[3] AFM straining,[2] and more. While loads and strains can easily be measured, one of the difficulties encountered in these tests is an accurate determination of the cross-sectional area of a nanotube[5] for the calculation of stresses. In this work we address a similar problem in the measurement of the properties of mats of aligned carbon nanotubes. It is important to understand the mechanical behavior of these mats, and accurate independent determination of the stiffness and moment of inertia of mats of nanotubes is very challenging, due to (a) varying tube-tube distances within the mat, (b) attractive forces between tubes due to van der Waals interactions (as high as nm force per nm CNT length),[6] and (c) sliding of tubes past each other within the mat. However, to determine the mechanical response of nanotubes for different loading conditions, such as deflection under shear ($\delta=PL^3/3EI$)[7] or buckling ($P_{cr}=\pi^2EI/L^2$),[7] only the flexural rigidity (=EI) is required. We demonstrate a simple optical method to observe the deflection of nanotubes and measure the flexural rigidity of mats of aligned nanotubes in several different configurations.

EXPERIMENTAL DETAILS

Mats of carbon nanotubes were grown using a thermal chemical vapor deposition (CVD) process. Nanotubes were grown using thin (5-15nm) iron films evaporated through TEM grids of varying grid dimensions to form catalyst patterns on quartz substrates. CVD growth was carried out at 900°C with benzene as a carbon source in an Ar:NH$_3$ atmosphere (roughly 3:2 mixing ratio). These patterned arrays were grown to heights ranging from 35µm to 60µm, as shown in Figure 1a.

The quartz substrate supporting the patterned nanotubes was placed inside the experimental apparatus, shown in Figure 1b. This equipment consisted of a quartz tube with an inner diameter of 6.2mm connected to a pressurized gas line (to introduce fluid flow). The sample was held in place by plastic inserts (to prevent sample vibration) and was exposed to flow across a range of velocities, controlled by adjusting the pressure (average fluid velocity was calibrated to gas pressure). In the experiment, the intensity of a polarized laser (λ=633nm) was monitored as it was passed through the nanotube mat. This laser was focused to a roughly 30µm spot size and positioned at the desired location on the sample (over the patterned region). This laser spot size was chosen to be small enough to resolve the average deflection behavior of a small group of CNTs. For the experiments, the laser beam was aligned parallel to nanotube axis and perpendicular to the substrate. A photo-detector was used to monitor the intensity change of the transmitted laser beam at different fluid velocities to demonstrate the bending of the CNTs.

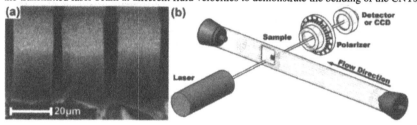

Figure 1: (a) Patterned arrays of CNTs, (b) Schematic of experimental flow chamber apparatus

DISCUSSION

Measurements of vertically aligned CNT arrays on quartz substrates (with known pattern geometries determined by the TEM grid used during catalyst patterning) showed that the change in transmitted laser intensity is in excellent agreement with the coverage of the CNTs, indicating CNTs absorb the laser light passing through the mats. As the fluid flow velocity in the tube is increased during the experiment, the transmitted laser intensity at CNTs spot was observed to decrease, as shown in Figure 2a. The reduction of transmitted intensity can be attributed to additional coverage from the nanotubes in the same laser spot, which is resulted from the deflection of the CNTs, and drops in intensities were recorded for every application of flow. The change in intensity for a given flow speed was consistent and a repeatable drop was observed for several consecutive measurements (cycling the flow from "off" to "on" at a given velocity to "off" again). The response to these velocity changes was also rapid, occurring immediately after either the beginning or cessation of the fluid flow (the loading or unloading of the CNTs).

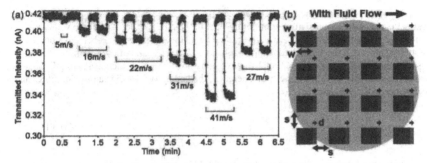

Figure 2: (a) Drops in transmitted laser intensity corresponding to applications of fluid flow, (b) Schematic of the laser spot over the patterned CNT arrays

The transmitted intensity in the absence of fluid flow is related to the area initially covered by CNT mats $(1 - A_{cnt}/A_{total})$ and is determined from the geometry of Figure 2b. When subject to fluid flow, the nanotubes deflect a distance d, covering an extra area ($w \times d$), where w is the width of the CNT pattern. The nanotube coverage can then be given by ($w^2 + d \times w$), and the normalized intensity is given by Equation 1, where s is the spacing between CNT arrays. Nanotube deflection can be calculated as a function of the CNT pattern geometry, the applied fluid velocity, and the fluid density. As the drop in intensity is proportional to the deflection and the drag force on the nanotubes, and the drag force is proportional to the velocity squared, an $I \propto V^2$ relationship is expected, and the observed intensity measurements fit this profile well.

$$\frac{I}{I_0} = \left(1 - \frac{w^2 + dw}{(w+s)^2}\right) \qquad (1)$$

Modeling and Simulation

In order to determine the flexural rigidity of these nanotube arrays, the fluid flow was modeled, and the corresponding drag forces and nanotube deflections were calculated. For the deflection modeling, the nanotubes were considered to act as fixed end cantilever beams,; by fitting the simulated deflections to the experimentally observed deflections, the flexural rigidity of these arrays was found.

In our system, there are three important length scales that must be considered when modeling the fluid flow. In each case fluid flow was evaluated through $Re = \rho U L/\mu = UL/\nu$, where ν, μ, and ρ are the kinematic and dynamic viscosities and density, respectively,[8] for either air or argon gases (the two fluids used during these experiments). For flow through the pipe (with characteristic length $L = 6.2$mm, the pipe diameter), the flow is found to be turbulent ($Re \approx 1$-2×10^4), and the peak velocity (in the tube center), is given by Equation 2.

$$U_{peak} = 1 + \frac{0.722}{\log(Re/6.9)} U_a \qquad (2)$$

209

In our experiments, the nanotube sample is located at the center of the tube on a flat substrate (see Figure 1b), and after considering flow in the pipe we then model the velocity profile by considering fluid flow over a flat plate. Due to the small distances between the nanotube sample and the leading edge of the substrate ($L \ll 1$mm), this flow is laminar ($Re \ll 10^6$), and the boundary layer height and velocity profile can be calculated using Equations 3 and 4, respectively (u is the velocity at a height y, the boundary layer δ is a function of Re and the distance to the substrate edge, x).

$$\frac{\delta}{x} = \frac{5}{Re^{1/2}}, \quad \frac{u}{U_{peak}} = 2\frac{y}{\delta} - (\frac{y}{\delta})^2 \qquad (3), (4)$$

Once the entire velocity profile along the CNTs' length is known, the drag forces must be calculated. Due to the small diameters of the CNTs used (~50nm in average), the Reynolds numbers for flow around these tubes were very small, and appropriate expressions for drag coefficients had to be used. For arrays of infinitely long, macroscopic cylinders, the Stokes-Oseen equation is used to predict the drag coefficient at low Reynolds numbers, and a correction for φ (the volume fraction occupied, 0.05 for our mats) can be applied.

In addition, the no-slip assumption of the conventional Navier-Stokes continuum mechanism (zero velocity right next to the surface), is no longer valid. With the size scales found for flow through mats of CNTs (for flow between CNTs, with an average spacing of roughly 150nm), a correction for this must be introduced. The Knudsen number (Kn) is the ratio of the mean free path of molecules in the fluid to the characteristic length of a system,[9] and in our system, the Knudsen number is roughly 0.5 (for both air and argon gases). As the Knudsen number must be smaller or equal to 10^{-3} for the no-slip condition to be valid, slip at surfaces must be considered, and this will reduce the effective drag on the object. With these considerations, a complete expression for the drag coefficient is obtained, including terms for the slip correction, drag on a single cylinder, and a correction for flow through a cylinder array (terms 1, 2, and 3 in Equation 5, respectively).

$$C_D = \left(\frac{1+4Kn}{1+6Kn}\right)\left(\frac{8\pi}{Re_{s-o}\ln(7.4/Re_{s-o})}\right)\left(\frac{3+2\phi^{5/3}}{3-4.5\phi^{1/3}+4.5\phi^{5/3}-3\phi^2}\right) \qquad (5)$$

Subsequent to the determination of C_D, the drag force per unit length, F_D, is found using equation 6, as a function of fluid velocity and density, (V and ρ), and the cross-sectional area, A (the product of CNT diameter, d, and step height, Δy).[8] The deflection (x) of carbon nanotubes is modeled by treating the tubes as fixed end cantilevers (Equation 7) and applied appropriate boundary conditions,[7] where $w(y)$ is the distributed force per unit length (from the drag forces).

$$F_D = \frac{1}{2}C_D A\rho V^2, \quad \frac{d^4x}{dy^4} = -\frac{w(y)}{EI} \qquad (6), (7)$$

Using the flow conditions and sample geometries for each sample tested, drag forces are determined and the flexural rigidity (EI) is then calculated by solving for the value that provides the best fit between the simulated and experimentally measured nanotube deflections. Several samples were studied in this way, covering a range of nanotube lengths (35-60μm) and fluid velocities (~5-65 m/s). The average flexural rigidity value was found to be ~7.9 x 10⁻¹⁶ Nm², with a relatively small spread in values from different samples (6.5-9.5 x 10⁻¹⁶ Nm²). Figure 3 shows experimental data points and the predicted deflections using the best fit flexural rigidity.

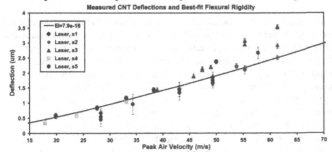

Figure 3: Deflection values obtained via transmitted laser intensity changes and predicted deflections using our fluid flow simulation and a flexural rigidity of 7.9 x 10⁻¹⁶ Nm²

There have been several other theoretical studies published on the flow of fluids through nanotube mats, using both non-equilibrium molecular dynamics,[10,11] and computational fluid dynamics.[12] We compared the drag forces calculated using our model (Equation 5) with the drag forces equations given in these publications (typically a function of Re and volume fraction φ, taking the form of $C_D=(a/Re)\varphi^b$), and obtained similar results, suggesting the CNT deflections are well approximated by the assumption of the Knudsen effect incorporated into a Stokes-Oseen model. In addition, a CCD-based experimental method (reported previously by us[13]) provided deflection measurements consistent with those obtained using this laser intensity procedure.

The drag forces for a given sample configuration are directly proportional to the fluid density (Equation 6), and a flow velocity-deflection relation made using one fluid can easily be calibrated for a different fluid using the ratio of the fluid densities. This was demonstrated by making deflection measurements using both air and argon gases on the same sample. A best fit velocity-deflection relation was determined using only the air flow data, and this relation was then corrected using the ratio of argon and air densities. Excellent agreement (Figure 4) is observed between the experimentally measured deflections (under argon flow) and the predicted deflections (from corresponding fluid velocities from the air flow data).

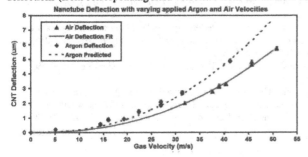

Figure 4: Agreement between measured argon deflection and deflection predicted using air flow results and a density correction

CONCLUSIONS

The flexural rigidity of carbon nanotube mats has not been widely studied to date, and in this work, we present an experimental measurement of this value. In addition, we demonstrated a simple method for monitoring the deflection of patterned arrays of carbon nanotubes subject to drag forces due to fluid flow. Fluid flow across the nanotubes was modeled, and numerical values for the flexural rigidity (EI), an important quantity when considering mechanical deformation such as buckling and deflections, were obtained. Consistent flexural rigidities (with an average value of 7.9×10^{-16} Nm^2) were obtained for a number of different nanotube samples over a range or measurement configurations. In addition, the optical method presented here would be suitable for mechanical characterization of carbon nanotubes and other nanostructures. With flow-deflection calibration, the techniques described in this work can be applied to a wide range of fluids with a simple density correction, and could also be adapted as the basis for the characterization of nano-scale flow, and for various applications, such as shear and tactile force sensors, micro- and nano-fluidics, and high sensitivity gas sensing, which has been demonstrated in this study.

REFERENCES

[1] S. Iijima, Nature **354**, 56-58 (1991).

[2] M.-F. Yu, O. Lourie, M. J. Dyer, K. Moloni, T. F. Kelly, and R. S. Ruoff, Science **287**, 637-640 (2000).

[3] B. I. Yakobson, C. J. Brabec, and J. Bernholc, Physical Review Letters **76**, 2511-2514 (1996).

[4] M. R. Falvo, G. J. Clary, R. M. Taylor II, V. Chi, F. P. Brooks Jr, S. Washburn, and R. Superfine, Nature **389**, 582-584 (1997).

[5] Y. Huang, J. Wu, and K. C. Hwang, Physical Review B **74**, 245413 (2006).

[6] M. R. Falvo, R. M. T. II, A. Helser, V. Chi, F. P. B. Jr, S. Washburn, and R. Superfine, Nature **397**, 236-238 (1999).

[7] F. P. Beer, J. E. Russell Johnson, and J. T. DeWolf, *Mechanics of Materials*, 4th ed. (McGraw Hill, Boston, 2006).

[8] R. W. Fox and A. T. McDonald, *Introduction to Fluid Mechanics*, 5th ed. (John Wiley & Sons, Inc., New York, 1998).

[9] R. W. Barber and D. R. Emerson, in *Proceedings of the 23rd international symposium on rarefied gas dynamics*; Vol. *663*, edited by A. D. Ketsdever and E. P. Muntz (AIP Press, Whistler, British Columia, 2003), p. 808-815.

[10] J. H. Walther, T. Werder, R. L. Jaffe, and P. Koumoutsakos, Physical Review E **69**, 062201 (2004).

[11] W. Tang and S. G. Advania, The Journal of Chemical Physics **236**, 174706 (2006).

[12] A. N. Ford and D. V. Papavassiliou, Ind. Eng. Chem. Res. **45**, 1797-1804 (2006).

[13] C. Ni, C. Deck, K. Vecchio, and P. R. Bandaru, Applied Physics Letters **92**, 173106 (2008).

Mechanical Properties and Energy Applications

Mater. Res. Soc. Symp. Proc. Vol. 1142 © 2009 Materials Research Society

Polymer Electrolyte Membrane Fuel Cell with Vertically Aligned Carbon Nanotube Electrode

Junbing Yang, Gabriel Goenaga, Ann Call and Di-Jia Liu*
Chemical Sciences & Engineering Division, Argonne National Laboratory,
9700 S. Cass Ave, Argonne, IL 60439, USA

ABSTRACT

Membrane electrode assemblies (MEA) using vertically aligned carbon nanotubes (ACNTs) as the electrocatalyst support for proton exchange membrane fuel cell (PEMFC) was developed through a multiple-step process, including chemical vapor deposition, catalyzing, fabricating MEA and packaging single cell. *I-V* polarization study demonstrated improved fuel utilization and higher power density in comparison with the conventional, ink based MEA.

INTRODUCTION

Carbon nanotubes (CNTs) have been considered as a promising electrode catalyst support material for proton-exchange membrane fuel cells [1]. The desirable attributes of CNTs include their unique geometric shape, high surface area, excellent thermal/electrical conductivities, and stability in the oxidative environment. There have been previous reports on CNT-based membrane electrode assemblies (MEAs) for fuel cells [2-4]. The CNTs in these electrodes are randomly oriented. The advantages associated with the distinctively structural properties of CNTs were not fully utilized. Reported here is our recent progress in fabricating and evaluating MEAs made of catalyst-decorated, vertically aligned carbon nanotubes (ACNTs). The potential advantages of ACNT-based MEAs include improved thermal and charge transfer through direct contact between the electrolyte and the current collectors and maximum exposure of the catalyst sites to the gas reactant through uniform support geometry and parallel alignment. Furthermore, a 3-D MEA with improved mass-transport, water management, and fuel utilization can be fabricated by patterning ACNT substrate with micrometer precision [5]. Our preliminary results demonstrate that such improvements are indeed achievable.

EXPERIMENT

ACNT-MEA fabrication

Our fabrication process of ACNT-based MEA involves three key steps: ACNT layer growth, catalyzing the ACNT, and transferring the catalyzed ACNT to the polymer electrolyte surface.

The ACNT layer is typically prepared by a chemical vapor deposition (CVD) process over a quartz substrate in a tubular reactor with two independently controlled temperature zones. The CVD precursors are vaporized or sublimated in the first, low temperature zone and ACNTs are grown over a substrate in the second, high temperature zone. The precursors typically consist of Fe-organometallic compounds either in solution, such as ferrocene dissolved in xylene, or as a solid, such as iron-phthalocyanine, mixed with carrier gas of Ar/H_2. The ACNT morphology depends strongly on the CVD reaction conditions and the type of precursors used. The ACNTs prepared in our laboratory typically have diameters ranging from 20 to 50 nm with adjustable lengths from 5 μm to >50 μm. The areal density of the tubes ranges from 10^8 to 10^9 tubes/cm^2, which results in a ~1000 to 4000-fold enhancement over the geometric surface area. A scanning electron microscopy (SEM) image of a typical ACNT layer is shown in Figure 1(a). When ammonia or other N-containing organic compounds were introduced into the CVD process, electrocatalytically active sites could also be formed on the surface of the ACNTs during their growth [6].

Figure 1. SEM images of (a) ACNT layer and (b) bundle cross-section prepared using ferrocene/xylene mixture.

The ACNT layer formed from CVD is subsequently catalyzed by a wet-chemical process. The extremely hydrophobic ACNT layer poses a significant challenge for the conventional aqueous catalyzing techniques. In our laboratory, we developed several approaches that mitigate the hydrophobicity and produce a dispersed catalyst layer using industrial Pt precursors such as $PtSO_3$, H_2PtCl_6, $(NH_3)_4Pt(NO_3)_2$, etc. These approaches generally involve altering catalyst solution composition so that it can easily infuse into the depth of the ACNT layer.

The catalyzed ACNT layer is subsequently transferred to a Nafion® membrane electrolyte surface using a hot-press method. We start by spray-coating a perfluorosulfonic acid ionomer solution to the catalyzed ACNT layer to ensure an uniform penetration of the ionomer solution into the ACNT bundles. To fabricate an ACNT-MEA, a Na^+ exchanged Nafion® 112 film is sandwiched between an ionomer-coated ACNT layer supported on a quartz plate (cathode) and an ink-based Pt/C decal (anode). The whole assembly is hot-pressed at 205°C and 120 lbs/cm^2 for 5 min. The MEA is formed after the quartz plate and the decal backing are separated. Shown in Figure 1(b) is an SEM image of an ACNT layer implanted on top of Nafion® membrane prepared by this method.

ACNT-PEMFC single cell test

The ACNT-MEAs thus prepared have an active area of 5 cm^2. They are subsequently mounted in a single cell test assembly with graphite bipolar plates engraved with a single-serpentine flow field. Commercial gas diffusion layers (GDL) made of woven PAN carbon fiber cloth coated with Teflon are inserted between the MEA and the bipolar plates. The single cell tests are conducted using hydrogen feed to the anode and air or oxygen to the cathode. Performance of the MEA is evaluated by measuring the current-voltage (I–V) polarization curves. Before each test, the MEA is first conditioned according to the US Fuel Cell Council test protocol [7]. After a stable performance is achieved, the polarization currents are measured by potentiostatically stepping the voltage between 0.2 volt and 1 volt. All tests are conducted at a cell temperature of 75°C, using 100% relative humidity for both air and hydrogen. The flow rates of anode and cathode gases are typically maintained at 100 sccm and 300 sccm at 1.2 and 1.5 bar, respectively. Following the polarization study, diagnostic tests, including impedance spectroscopy and cyclic voltammetry are performed to characterize the cell ohmic resistance and electrocatalytic surface area. For comparison, a commercial MEA prepared by the ink-based process is also tested under the similar conditions. High resolution SEM and Energy-Dispersive X-ray Spectrometry (EDS) are used to probe the morphology and to determine the Pt loading.

DISCUSSION

Potential advantages of ACNT-MEA

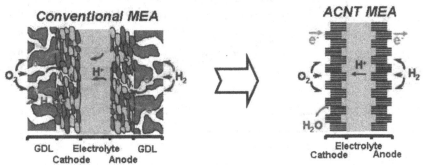

Figure 2. Schematic drawings of conventional versus ACNT based MEAs.

The catalytic electrodes in a conventional MEA are typically fabricated by an ink casting process [8]. A schematic representation is shown on the left side of Figure 2. In such a process, the ionomer has to perform dual tasks, conducting the protons and "gluing" together the catalyst particles, which could compromise the MEA performance. For example, excess ionomer may over cast the polymer to partially block reactant passage to the active sites, whereas insufficient ionomer could lead to poor proton transfer and a weakened adhesion of the catalyst layer. Another drawback of the ink process arises from the morphology of the resulting carbon matrix. Carbon particles are randomly packed. The electronic and thermal conduction are accomplished through the percolation between the particle contacts. If such contact is interrupted, for example through the shrinkage of the carbon particles from oxidation at high polarization potentials under fuel cell operations, both conduction as well as the catalytic reaction will be impeded.

An ACNT-based MEA offers the solution to these issues. A schematic drawing is shown on the right side of Figure 2. In the ACNT-MEA, the surface of the nanotube is decorated with the catalytic sites which are exposed to the gas reactant during the electrochemical reaction. Electron transfer is through the graphene layer of the nanotube. The proton transfer can be accomplished by a very thin, gas-permeable ion-conducting coating prepared with a diluted Nafion® ionomer solution [9, 10]. Unlike the conventional MEA, the ACNT-MEA connects the membrane electrolyte and the current collector directly through the vertically oriented nanotubes "spring-loaded" through compression during the fuel cell packaging. Heat and electron transfer from the reaction sites can be accomplished effectively through individual graphitic carbon nanotubes, and the loss of efficiency through inter-particle percolation in a conventional MEA is eliminated. Direct contact by ACNT also minimizes the risk of conductivity degradation due to loss of carbon surface under the oxidative environment. An important benefit of the ACNT-MEA is the better reactant utilization through improved mass transfer. With the catalyst-decorated surface fully exposed through the ordered nanostructure, the reactant gas can reach the catalytically active sites with less resistance than the tortuous path in ink-based MEA. It is anticipated, therefore, that a lower mass-transport overpotential would be achievable over the high current region during the polarization.

Single cell test results and comparison

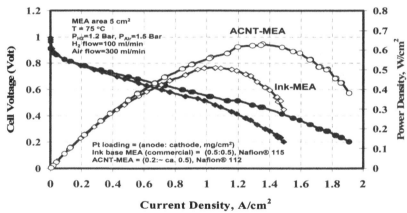

Figure 3. The cell voltages (solid symbols) and the power densities (hollow symbols) as functions of current in single-cell tests of an ACNT-MEA (circles) and a commercial MEA (diamonds).

Shown in Figure 3 are the I-V polarization and power density curves for an ACNT-MEA and an ink-based MEA, the latter procured from a commercial source. This particular ACNT-MEA has a carbon nanotube cathode of a thickness approximately 25 μm. A H_2PtCl_6 solution was used as the platinum precursor to catalyze the ACNT layer with the cathode Pt loading of approximately 0.5 mg/cm². The anode is prepared by the ink-process, using 20 wt% Pt/carbon (Vulcan XC72) with a Pt loading of 0.2 mg/cm². For the commercial MEA, the Pt loadings are 0.5 mg/cm² for both cathode and anode. The electrolyte was Nafion® 115.

In the kinetically limited region, little difference is observed between ACNT-MEA and the commercial product. In the high current region, the I-V curve extends further for ACNT-MEA, suggesting a significant reduction of the overpotential from the mass-transport limit. Correspondingly, the power density improves in the high current region. This observation demonstrates experimentally that improved interaction between the reactant gases and the catalyst sites can indeed be achieved by the ACNT electrode structure without flooding. In addition to polarization curves, we also investigated other MEA properties such as impedance and intrinsic electrochemical surface area (ECA), $A_{Pt,MEA(ca)}$. The ACNT-MEA generally exhibits low impedance under normal operation, although the portion of the reduction through the carbon nanotubes is unclear since the major contribution is from the membrane. We also measured the polarization with oxygen feed and obtained the mass activity, $i_{m(0.9\ V)}$ and the specific activity, $i_{s(0.9\ V)}$ at 0.9 volts for several MEAs using IR corrected Tafel-lines [11]. Some of these parameters are listed in Table I.

TABLE I. Electrocatalyst parameters of selected MEA samples

MEA	Commercial MEA	ACNT-MEA-1	ACNT-MEA-2
Anode/Cathode (mg$_{Pt}$/cm²)	0.5/0.5	0.2/0.4	0.2/0.4
$A_{Pt,MEA(ca)}$ (m²/g$_{Pt}$)	34	27	27
$i_{m(0.9\ V)}$ (mA/mg$_{Pt}$)	94	50	101
$i_{s(0.9\ V)}$ (µA/cm²$_{Pt}$)	276	185	374

We should point out that, although the preliminary improvements are observed, a significant amount of study remains to be carried out to fully utilize the unique electrode structure of ACNT. For example, we believe that platinum loading can be further reduced through optimizing the ACNT nanostructure and the catalyst dispersion. We need also to identify the optimal ACNT morphology (tube density, length, etc.) that can provide the ideal balance between catalyst distribution, mass transfer, and water management. For example, we found in a separate experiment that proton transfer can occur over bare carbon nanotube surface, although the transport efficiency appears to be significantly lower than that of Nafion® ionomer coating, suggesting additional experimental parameter to be optimized.

CONCLUSIONS

ACNT based MEAs were successfully fabricated through a series of material and process development. The new MEAs were evaluated in a PEMFC single cell test stand. The preliminary results demonstrated improved power density at high current region over that of conventional, ink based MEA, supporting the hypothesis that better mass transport and thermal/electrical conductivities can be achieved through the unique nanostructure of ACNTs.

ACKNOWLEDGMENTS

This work was supported by the U.S. Department of Energy, Office of Energy Efficiency and Renewable Energy, Office of Hydrogen, Fuel Cells, and Infrastructure Technologies. The electron microscopy was performed at the Electron Microscopy Center for Materials Research at Argonne National Laboratory, a U.S. Department of Energy Office of Science Laboratory operated under Contract No. DE-AC02-06CH11357 by UChicago Argonne, LLC. The authors are grateful to Drs. N. Kariuki and D. Myers for their experimental support and helpful discussions. G. Goenaga would like to express his gratitude for the partial financial support from Professor S. Greenbaum of Hunter College, City University of New York.

REFERENCES

1. Y. Gogotsi, *Nanotubes and Nanofibers*. CRC Press, Boca Raton. 2006.
2. Y. H. Liu, B. Yi, Z. G. Shao, L. Wang, D. Xing, H. Zhang, *Journal of Power Sources*, **163**(2) 807-813 (2007)
3. W. Li, X. Wang, Z. Chen, M. Waje, Y. Yan, *Langmuir* **21**(21), 9386-9389 (2005)
4. J. M. Tang, K. Jensen, M. Waje, W. Li, P. Larsen, K. Pauley, Z. Chen, P. Ramesh, M. E. Itkis, Y. Yan, and R. C. Haddon, *J. Phys. Chem. C* **111**, 17901-17904 (2007)
5. J. Yang and D.-J. Liu, *Carbon* **45**, 2843–2854 (2007)
6. J. Yang, D.-J. Liu, N. Kariuki and L. X. Chen, *Chem. Comm.* **3**, 329 – 331 (2008)
7. http://www.usfcc.com
8. M. S. Wilson and S. Gottesfeld, *J. of Appl. Electrochem.* **22**, 1 (1992)
9. R. J. Mashl, S. Joseph, N. R. Aluru and E. Jakobsson, *Nanotech.* **1**, 152 - 153 (2003)
10. D. J. Mann and M. D. Halls, *Phys. Rev. Lett.* **90**(19), 195503 (2003)
11. H. A. Gasteiger, S. S. Kocha, B. Sompalli and F. T. Wagner, *Appl. Cataly.* **56**, 9-35 (2005)

Mater. Res. Soc. Symp. Proc. Vol. 1142 © 2009 Materials Research Society   1142-JJ20-20

Control of NEMS Based on Carbon Nanotube: Molecular Dynamics Study

Irina V. Lebedeva[1,2,3], Andrey A. Knizhnik[1,2], Olga V. Ershova[3], Andrey M. Popov[4], Yurii E. Lozovik[4], and Boris V. Potapkin[1,2]
[1]Kintech Lab Ltd, Kurchatov Sq., 1, Moscow, 123182, Russia
[2]RRC "Kurchatov Institute", Kurchatov Sq., 1, Moscow, 123182, Russia
[3]Moscow Institute of Physics and Technology, Institutskii pereulok, 9, Dolgoprudny, Moscow Region, 141701, Russia
[4]Institute of Spectroscopy, Troitsk, Moscow Region, 142190, Russia

ABSTRACT

Molecular dynamics simulations of nanotube-based nanoelectromechanical systems (NEMS) controlled by a non-uniform electric field are performed by the example of the gigahertz oscillator. The amplitude of the control force needed to sustain the oscillation is found to increase with temperature. It is shown that thermodynamic fluctuations lead to the oscillation breakdown and restrict the sizes of the NEMS for which the control of the NEMS operation is possible.

INTRODUCTION

Carbon nanotubes are considered to be a promising material for the use in nanoelectromechanical systems (NEMS). Ability of free relative motion of the walls of carbon nanotubes [1, 2] and their excellent wearproof characteristics [2] allow using carbon nanotube walls as movable elements [3–7]. However, the possibility to control the operation of nanotube-based NEMS requires further investigation. Here we examine one of the methods for controlling the motion of NEMS based on carbon nanotubes which was proposed recently [8]. In this method a functionalized wall with an electrical dipole moment can be controlled by a non-uniform electric field.

As an example of nanotube-based NEMS we consider the gigahertz oscillator based on a double-walled carbon nanotube (DWNT, see figure 1a), which is commonly used as a model system to study the behavior of nanotube-based NEMS [9–17]. Upon the telescopic extension of the inner wall outside the outer wall, the van der Waals force F_W retracts the inner wall back into the outer wall, thereby makes this NEMS oscillate. However, these oscillations are non-harmonic and dissipative with the Q-factor $Q \approx 10 - 1000$ [9–17] (the Q-factor of the system is defined as a ratio of the total oscillation energy to the energy loss per one oscillation period). Since the frequency of the damping oscillations increases with decreasing the oscillation amplitude [3, 4], this frequency increases with time [10]. To provide the stationary operation of the gigahertz oscillator, that is to keep its frequency constant, it is necessary to compensate the energy dissipation by the work of an external force.

The possibility to control a functionalized nanotube wall by a non-uniform electric field is studied here using molecular dynamics (MD) simulations of the controlled operation of the gigahertz oscillator. To estimate the amplitude of the control force required to sustain the oscillations we calculate the Q-factor of the gigahertz oscillator on the basis of MD simulations of the damping oscillations. These simulations also reveal significant thermodynamic fluctuations in the NEMS under consideration.

METHODS

We considered the $(5,5)@(10,10)$ DWNT-based oscillator with the both nanotube walls equal in length [8] (see figure 1a). The inner wall had one end capped and the other end open and terminated with 10 hydrogen atoms. Both ends of the outer wall were open and not functionalized.

The analysis of the free and controlled behavior of the oscillator was performed using empirical interatomic potentials. Van der Waals interaction between the inner and outer wall atoms was described by the Lennard–Jones 12–6 potential

$$U = 4\varepsilon \left(\left(\frac{\sigma}{r} \right)^{12} - \left(\frac{\sigma}{r} \right)^{6} \right) \tag{1}$$

with the parameters $\varepsilon_{CC} = 3.73$ meV, $\sigma_{CC} = 3.40$ Å and $\varepsilon_{CH} = 0.65$ meV, $\sigma_{CH} = 2.59$ Å for carbon-carbon and carbon-hydrogen interaction, respectively, obtained from the AMBER database [18] for aromatic carbon and hydrogen bonded to aromatic carbon. The parameters provide a consistent description of the pairwise carbon-carbon and carbon-hydrogen interactions. Moreover, these parameters were shown to provide the average value of the van der Waals force which retracts the telescopically extended inner wall back into the outer wall $F_W \approx 1200$ pN, in good agreement with the experiment [2]. The cut-off distance of the Lennard–Jones potential was taken to be 12 Å. The covalent carbon-carbon and carbon-hydrogen interactions inside the walls were described by the empirical Brenner potential [19].

An in-house MD-kMC code was implemented. The code used the velocity Verlet algorithm and neighbor lists to improve the computing performance. The time step was 0.2 fs, which is about 100 times smaller than the period of thermal vibrations of hydrogen atoms. At the beginning of the MD simulations, the inner wall was pulled out along the DWNT axis by about 30% of its length and released with zero initial velocity. The outer wall was fixed at three atoms. The relative fluctuations of the total energy of the system caused by numerical errors were less than 0.3% of the interwall van der Waals energy.

RESULTS AND DISCUSSION

In the case of a relatively long DWNT and a large oscillation amplitude, one can neglect the small telescopic extensions of the inner wall (< 0.5 nm) where the interwall van der Waals force increases from zero to the nearly constant value F_W. Under this assumption, the critical amplitude F_{0c} (i.e. the minimal value at which the work of the external force can compensate the energy dissipation) of the harmonic control force $F(t) = F_0 \cos \omega t$, where ω is the frequency of the stationary oscillation, is given by [8]

$$F_{0c} = \frac{\pi^2 F_W}{32Q}.$$ (2)

So, to estimate the amplitude of the control force needed to sustain the oscillations, one should know the oscillator Q-factor. We calculated the dissipation rate for different temperatures and DWNT lengths by means of the microcanonical MD simulations of free oscillations (see figure 1b). These data were used to estimate the Q-factor. We took a simulation time of 500 ps, which minimizes the Q-factor calculation error in a single run and provides the accuracy of the Q-factor calculations within 20%. The temperature change over the simulation time was less than 9% at pre-heating temperatures of 50, 100, 150 and 300 K. At a pre-heating temperature of 0 K, the temperature increased to 2 K within the simulation time.

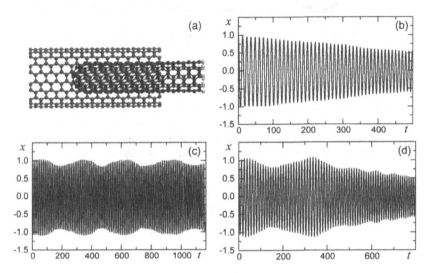

Figure 1. (a) Structure of the (5,5)@(10,10) DWNT-based oscillator. Hydrogen atoms are shown in light grey. (b-d) Calculated displacement of the movable wall x (in nm) as a function of time t (in ps) at temperature of 300 K. (b) Free oscillation. (c, d) Controlled oscillations for the amplitude of the voltage of (c) 60.5 V and (d) 61.4 V. The DWNT length is 3.1 nm.

The calculated Q-factors Q for different temperatures and DWNT lengths are given in tables I and II. A considerable decrease in the Q-factor is found for the DWNT that is shorter than 3 nm. For the longer DWNTs, the Q-factor is almost independent of the length if the ratio of the oscillation amplitude to the DWNT length is maintained, which conforms to [13, 14]. The Q-factor strongly increases with a decrease in temperature, in agreement with [15–17]. Using the obtained Q-factors, we estimated the critical amplitudes of the control force at different temperatures according to equation 2 (see table I).

In order to demonstrate the possibility of controlling the motion of a nanotube wall with a dipole moment by a non-uniform electric field, we performed the MD simulations of the gigahertz oscillator with the both walls 3.1 nm in length exposed to the harmonic electric field of the spherical capacitor with the radii of the plates of 100 and 110 nm. The considered inner wall (with one end capped and the other end open and terminated with hydrogen atoms, see figure 1a) possesses the dipole moment of $4.5 \cdot 10^{-29}$ C·m [8]. The charge distribution in the inner wall was taken in accordance with the calculations performed in [8]. The frequency of the electric field corresponded to the initial oscillation frequency. The initial phase shift between the electric field and inner wall velocity equaled zero. In these calculations, the outer wall temperature was maintained by a periodic rescaling of atomic velocities every 0.1 ps (the Berendsen thermostat [20]).

Table I. Calculated Q-factor Q, critical amplitude of the control force F_{0c}, critical amplitude of the voltage U_{0c} and relative deviation δ of the energy change over a half-period of the oscillations at different temperatures T for the 3.1 nm DWNT.

T (K)	Q	F_{0c} (pN)	U_{0c} (V)	δ
0	700 ± 350	0.52	6.0	4.4
50	230 ± 50	1.4	17	1.2
100	170 ± 30	2.3	26	1.4
150	100 ± 20	2.7	31	1.5
300	48 ± 10	6.6	77	1.5

Table II. Calculated Q-factor Q and relative deviation δ of the energy change over a half-period of the oscillations for different DWNT lengths L at temperature 150 K.

L (nm)	Q	δ
2.4	67 ± 13	1.2
3.1	120 ± 20	1.3
3.8	100 ± 20	1.2
4.6	120 ± 20	0.9
6.3	110 ± 20	0.7

The results of the MD simulations of the controlled oscillations are presented in figures 1c,d. The amplitude of the applied voltage was 60.5 and 61.4 V at temperature of 300 K. As seen from figure 1c, the oscillation can be sustained at a constant amplitude with the amplitude of the voltage which is less than that corresponding to equation 2 (see table I). This discrepancy can be explained in the following way. Equation 2 for the critical amplitude of the control force is derived for a long oscillator with a large oscillation amplitude assuming that the interwall van der Waals force F_W does not depend on the telescopic extension of the inner wall. However, for the oscillator of 3.1 nm length, the interwall van der Waals force is not constant and has a smaller average value. The variation of the oscillation amplitude in figure 1c can be related to the fact that the amplitude of the control force exceeds the critical value.

Note that the considered voltages are of the order of the voltages (up to 100 V) reached in a nanomotor [5] which was implemented recently. Furthermore, the amplitude of the applied voltage needed to sustain the oscillation at a constant amplitude can be decreased if the inner wall has a higher dipole moment. This may be achieved, for example, by the adsorption of hydrogen and fluorine atoms at opposite open ends of the wall [8]. Moreover, as shown above, the Q-factor strongly increases with decreasing temperature. Therefore, the oscillator operation at low temperature requires a smaller amplitude of the control force. For instance, at liquid helium temperature of 4.2 K, the critical amplitude of the applied voltage is only several volts.

It should be mentioned that in some cases, even if the amplitude of the control force is relatively high, the breakdown of the oscillation occurs (see figure 1d). We suppose that this breakdown is induced by thermodynamic fluctuations. To estimate the level of the fluctuations in the system we calculated the relative energy change over every half period of the oscillations $\Delta E / E$ (see figure 2). Significant fluctuations of this quantity are observed. The calculated relative root-mean-square deviation δ of $\Delta E / E$ for different temperatures and DWNT lengths are given in tables I and II. As seen from tables I and II, $\delta > 1$ for the oscillator 3.1 nm in length. The relative deviation δ decreases with increasing the DWNT length and weakly depends on temperature under moderate conditions. However, the considerable increase in δ is observed for zero pre-heating temperature. Possibly this is related to the fact that the system is highly non-equilibrium at such a low temperature.

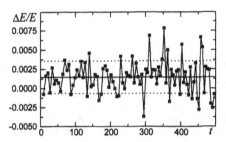

Figure 2. Calculated relative energy change $\Delta E / E$ over a half period of the oscillation as a function of time t (in ps) for the 3.1 nm DWNT at temperature of 300 K. The average value and the root-mean-square deviation are shown with the solid and dotted lines, respectively.

We assume that the oscillation breakdown can be explained in the following way. Suppose that occasionally a significant negative fluctuation of the oscillation energy occurs. This means that the oscillation amplitude decreases. Since the oscillator frequency strongly depends on the oscillation amplitude [3, 4], the oscillator gets out of the resonance with the control force. This leads to a decrease of the work of the control force, which, in turn, results in a further decrease of the oscillation energy. As follows from this explanation, the stability of the oscillator operation might be improved with increasing the amplitude of the control force. Since the magnitude of the fluctuations decreases with increasing the DWNT length (see table II), the stability of the oscillator operation can be also achieved by increasing its size. In other words, the fluctuations restrict the sizes of the NEMS for which the control of the NEMS operation is possible.

CONCLUSIONS

We performed molecular dynamics simulations of the controlled operation of the DWNT-based gigahertz oscillator which demonstrated the possibility to control the motion of a functionalized nanotube wall by a non-uniform electric field. The critical amplitude of the control force needed to sustain the oscillation was found to increase with temperature. Significant thermodynamic fluctuations were revealed in the NEMS. It was found that these fluctuations result in the oscillation breakdown. The fluctuations were shown to restrict the sizes of the NEMS for which the control of the NEMS operation is possible.

ACKNOWLEDGMENTS

This study was partially supported by the Belarussian-Russian Foundation for Basic Research within the framework of the joint program BRFFI-RFFI (project no. 08-02-90049-Bel) and the Russian Foundation for Basic Research (project no. 08-02-00685).

REFERENCES

1. J. Cumings and A. Zettl, *Science* **289**, 602 (2000).
2. A. Kis, K. Jensen, S. Aloni, W. Mickelson and A. Zettl, *Phys. Rev. Lett.* **97**, 025501 (2006).
3. Q. Zheng and Q. Jiang, *Phys. Rev. Lett.* **88**, 045503 (2002).
4. Q. Zheng, J. Z. Liu and Q. Jiang, *Phys. Rev. B* **65**, 245409 (2002).
5. B. Bourlon, D. C. Glatti, L. Forro and A. Bachtold, *Nano Lett.* **4**, 709 (2004).
6. Yu. E. Lozovik and A. M. Popov, *Phys. Usp.* **50**, 749 (2007).
7. A. M. Popov, E. Bichoutskaia, Yu. E. Lozovik and A. S. Kulish, *Phys. Stat. Sol. (a)* **204**, 1911 (2007).
8. O. V. Ershova, Yu. E. Lozovik, A. M. Popov, O. N. Bubel', N. A. Poklonski and E. F. Kislyakov, *Physics of the Solid State* **49**, 2010 (2007).
9. Y. Zhao, C.-C. Ma, G. Chen and Q. Jiang, *Phys. Rev. Lett.* **91**, 175504 (2003).
10. W. Guo, Y. Guo, H. Gao, Q. Zheng and W. Zhong, *Phys. Rev. Lett.* **91**, 125501 (2003).
11. W. Guo, W. Zhong, Y. Dai and S. Li, *Phys. Rev. B* **72**, 075409 (2005).
12. P. Liu, H. J. Gao and Y. W. Zhang, *Appl. Phys. Lett.* **93**, 083107 (2008).
13. J. L. Rivera, C. McCabe and P. T. Cummings, *Nano Lett.* **3** 1001 (2003).
14. J. L. Rivera, C. McCabe and P. T. Cummings, *Nanotechnology* **16**, 186 (2005).
15. C.-C. Ma, Y. Zhao, Y.-C. Yam, G. Chen and Q. Jiang, *Nanotechnology* **16**, 1253 (2005).
16. S. Xiao, D. R. Andersen, R. P. Han and W. Hou, *Journal of Computational and Theoretical Nanoscience* **3**, 142 (2006).
17. J. Servantie and P. Gaspard, *Phys. Rev. Lett.* **91**, 185503 (2003).
18. http://amber.scripps.edu//#ff.
19. D. W. Brenner, O. A. Shenderova, J. A. Harrison, S. J. Stuart, B. Ni and S. B. Sinnott, *J. Phys.: Condens. Matter* **14**, 783 (2002).
20. H. J. C. Berendsen, J. P. M. Postma, W. F. Gunsteren, A. DiNola and J. R. Haak, *J. Chem. Phys.* **81**, 3684 (1984).

Mater. Res. Soc. Symp. Proc. Vol. 1142 © 2009 Materials Research Society   1142-JJ20-44

Electronic Detection and Influence of Environmental Factors on Conductivity of Single DNA Molecule Using Single-walled Carbon Nanotube Electrodes

Harindra Vedala[1], Taehyung Kim[1], Wonbong Choi[1]
Sookhyun Hwang[2], Minhyon Jeon[2]

[1]Mechanical and Materials Engineering, Florida International University, Miami, FL, 33174 USA
[2]Department of Nano Systems Engineering, Inje University, Gimhae , 621-749, Korea

ABSTRACT

Rapid detection of ultra low concentration or even single DNA molecules are essential for medical diagnosis and treatment, pharmaceutical applications, gene sequencing as well as forensic analysis. In this work we demonstrate the use of single-walled carbon nanotubes as nanoscale electrodes for electronic detection of single DNA molecules. A detailed study on electrical conductivity and influence of environmental factors on a double-stranded DNA molecule bridging a single-walled carbon nanotube (SWNT) gap is presented. The amine terminated DNA molecule was trapped between carboxyl functionalized SWNT electrodes by dielectrophoresis. Typically, a current of tens of picoamperes at 1 V was observed at ambient conditions, with a decrease in conductance of about 30% in high vacuum conditions. As the intrinsic conductivity of the DNA molecule is strongly influenced by environmental factors we study the conductivity of DNA under the influence of counterion variation, salt concentration, and pH. The counterion variation was analyzed by changing the buffer type. A reversible shift in the current signal was observed for pH variation.

INTRODUCTION

The enormous amount of genetic information brought by extensive genome sequencing has raised the need for simple, fast, cheap and high-throughput miniaturized and mass-producible analytical devices to attend the growing market of molecular diagnostics. Recently direct electrical characterization of DNA has received great attention due to its significance in various fields ranging from biology to electronics [1-3]. It is believed that development of DNA detection systems based on electronic transduction will enable us to develop highly selective and ultrasensitive biosensors. Along with the biological relevance for understanding the mechanism of DNA oxidative damage, charge transport (CT) studies play an important role in technological development of DNA microarrays, molecular switches and biosensors[4]. Due to the highly dynamic structure of DNA, the CT is greatly influenced by its sequence, length and environmental conditions, yielding several controversial results ranging from insulator, semiconductor to even superconductor [5, 6]. However experimental investigations based on electrical and electrochemical techniques as well as theoretical studies have partially solved the controversy of CT through DNA[7]. It has been shown that CT in DNA takes place via π stacking of electron orbitals of the neighboring bases in a DNA double helix. For shorter distances the charge (hole) moves from one guanine-cytosine (GC) pair to another by coherent tunneling (superexchange mechanism) whereas for longer distances thermal hopping occurs [8]. While most of these results are based on indirect measurement techniques, direct electrical characterization of a single DNA molecule instead of a ensemble of molecules, would provide greater insight in understanding the CT through this promising biopolymer. From technological point of view, an electronic DNA detection platform could help in eliminating the need for amplification techniques such as polymerase chain reaction which is currently used for DNA sequencing and other related applications.

However, there exist several challenges in performing direct measurements of CT through a short DNA molecule (<40 nm) [9], which include its precise manipulation, reliable interfacing with the electrodes, electrode properties, DNA-substrate interaction and more importantly the environmental effects. Since the last decade several studies have been performed for understanding electrical conductivity of DNA molecules of various sequences and lengths ranging from a few nanometers to micrometers. A typical measurement involves attaching the DNA molecules between metal electrode such as gold, either by direct deposition from the solution or by using electric field to trap and align. As the width of the electrode is much larger than the diameter of DNA, the quantitative determination of the number of molecules attached between the electrodes is difficult. As a result, a large variation in the conductivity of DNA is observed in the literature. In this respect single-walled carbon nanotube (SWNT) electrodes are a promising alternative for electrical characterization of single molecules. For example SWNT as nanoelectrodes, with their nanoscale diameter (~ 2 nm) and excellent electronic and chemical properties, can be an ideal candidate for electrical characterization of DNA at a single molecular level. Few reports on the use of carbon nanotubes as nanoelectrodes were demonstrated for electrical characterization of various organic molecules [10,11]. We have recently reported the direct electrical measurement of a single hybridized DNA molecule using a pair of SWNT nanoelectrodes [12].

In this work, we report the direct electrical characterization of single DNA molecules and also study the influence of local environmental factors on their CT. Several earlier reports have shown the strong influence of humidity on the conductivity of DNA molecules[4,13]. However other environmental factors such as counterion variation, pH and temperature which have substantial effect on DNA conductivity were not studied in detail. Here we investigated the influence of water molecules on DNA conductivity by comparing current-voltage (I-V) characteristics measured in ambient and in high vacuum conditions. The effect of ionic variation on the electrical conductivity of dry DNA molecule was studied by measuring I-V characteristics of dsDNA previously suspended either in sodium acetate (NaAc) or Tris(hydroxymethyl) aminomethane (TRIS). In the case of sodium acetate the counterions were Na^+ while in TRIS buffer they were H^+ and $NH(CH_2OH)_3^+$. The pH and temperature effect on the CT of the trapped DNA molecule was also studied.

EXPERIMENT

Single-walled carbon nanotubes were synthesized by chemical vapor deposition technique and were suspended in isopropyl alcohol by ultrasonication. Initially a 2 μl droplet of SWNT suspension was spun on an thermally oxidized (500 nm) silicon substrate having photolithographically patterned microelectrodes and bonding pads [14]. Electrical contacts to individual SWNT were made by first locating them with respect to prepatterned index marks using field emission scanning electron microscope (FESEM) imaging. That was followed by making contact leads using e-beam lithography and sputtering of 50 nm of Au on 10 nm Ti adhesion layer. To fabricate a pair of nanoelectrodes, Focused Ion Beam (FIB) was used for etching near the center of an individual SWNT segment between the metal electrodes. FIB etching parameters (beam current, exposure time) were optimized to obtain a uniform gap in accordance with the length of the DNA strands. Figure 1 (a) shows a schematic illustration of the device fabrication process

Electrical conductivity of an 80 base–pair (bp) dsDNA fragment (contour length ~27nm), encoding a portion of the *H5N1* gene of avian *Influenza A* virus (AIV) was measured. The template strand obtained with amine modifications at the 5′ and 3′ ends was hybridized with the unmodified complementary strand in 10 mM NaAc buffer (pH 5.8) at equimolar concentrations. Hybridization was performed by heating the solution to 90°C for 5 min followed by a two degree decrease every one minute and held for 15 min at 25°C, using a thermocycler. A similar hybridization protocol was followed to prepare dsDNA molecules suspended in TRIS buffer [10 mM tris-HCl / 1.0 mM EDTA (pH 8.0)].

To measure the electrical conductivity of the dsDNA molecules, the SWNT nanoelectrodes were first functionalized with COOH groups to form a strong covalent bond with amine terminated DNA molecule using the procedure described by Chen et. al [15]. The sample was then incubated for 30 min in 2 mM 1-ethyl-3-(3-dimethylaminopropyl) carbodiimide hydrochloride and 5 mM *N*-hydroxysuccinimide (NHS) to convert carboxyl groups to amine-reactive Sulfo-NHS esters. Amine terminated dsDNA molecules from a diluted solution (10 nM) were deposited on the electrodes and a.c. dielectrophoresis technique was used to align and immobilize DNA molecule between the electrodes. The devices were then washed with corresponding buffer solutions to remove non specifically attached DNA molecules. The samples were blow dried with nitrogen stream [16]. Dielectrophoresis parameters namely frequency (0.1-10 MHz), voltage (0.1 - 5V$_{PP}$) and time (10 - 500 sec) were experimentally optimized. Control experiments such as conductivity measurement of DNA free buffers, affect of DNAse I (digestion enzyme), on DNA conductivity were performed. All the *I-V* measurements were carried out using four-probe station equipped with precision semiconductor parameter analyzer (Agilent 4156C) at ambient and in high vacuum (10^{-5} Torr) conditions. All measurements were made at room temperature (23°C).

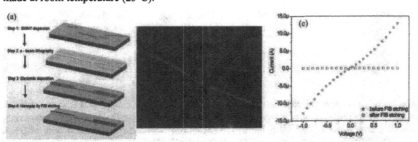

Figure 1.(a) Schematic of SWNT nanoelectrodes fabrication (b) AFM image of SWNT nanoelectrodes with a nanogap of 30 nm, scale bar 50 nm (c) *I-V* characteristics of SWNT before and after making a nanogap using focused ion beam etching process.

DISCUSSION

Each chip consisted of several SWNTs nanoelectrodes which were contacted by Ti/Au electrodes to micro pads. Figure 1 (b) shows the AFM image of nanoelectrodes fabricated using the FIB etching process. Before the formation of a nanogap, the I-V characteristics of each SWNT were measured. The SWNTs which exhibited metallic behavior were selected for the fabrication of nanogaps. Figure 1 (c) depicts I-V characteristics of a typical SWNT before and after the etching process. The SWNT exhibited resistance in the range of kΩ. After FIB etching the current decreased from several micro-amperes to a few femto-amperes (noise range of instrument) indicating the nanogap formation. FIB etching also resulted in the formation of a trench in the oxide layer beneath the gap.

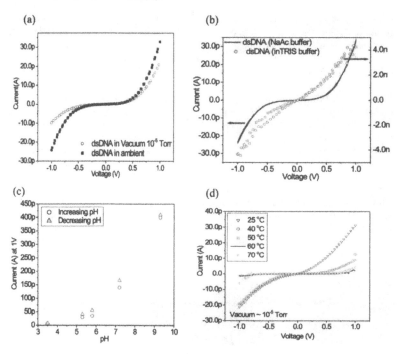

Figure 2. I-V characteristics of single dsDNA molecule measured using SWNT nanoelectrodes under ambient and high vacuum conditions (b) IV characteristics of dsDNA molecule previously suspended either in NaAc and TRIS buffer (c) Effect of pH variation on conductivity of DNA measured at 1V bias. I-V of the same device washed with buffer solution of different pH values is shown both for gradual increase and decrease in pH (d) Effect of temperature variation on DNA conductivity measured under high vacuum condition. A gradual decrease in electrical signal with increase in temperature was observed.

We used a.c. dielectrophoresis to trap and align the DNA molecule between the SWNT nanoelectrodes with a peak-to-peak voltage of 1 V and for a frequency range between 100kHz - 10MHz for 400 sec. The applied field (40 MV/m) is sufficiently high to overcome Brownian motion, which is dominant in nanoscale objects. The I-V characteristics of the double helix, hybridized in NaAc buffer, dried and measured at ambient and in high vacuum (10^{-5} torr) conditions is shown in figure 2 (a). At ambient conditions a current signal of 30 pA for a 1 V bias was observed while at high vacuum condition the signal decreased by 33 %. This decrease may be ascribed to the partial removal of water molecules from the proximity of the DNA. In fact, in ambient condition the proton-transfer process in the hydration layer surrounding the DNA promotes the electrical conductivity but diminishes in high vacuum [17].

Figure 2(b) shows a comparison of I-V characteristics of hybridized DNA (in dry state) previously suspended either in NaAc or TRIS buffer. A current signal in the range of 25-50 pA (at 1 V bias) was observed for the DNA molecule in the case of sodium acetate buffer. A nonlinear I-V characteristic was observed indicating a semiconducting behavior of the trapped DNA fragment encoding a specific gene but devoid of any periodic arrangement of the base pairs. In comparison, we observed that dsDNA in TRIS buffer exhibit almost two orders of magnitude higher current. To further investigate the cause of differences in magnitude, control experiments for both the buffer solutions were performed. To investigate the influence of buffer on the DNA conductivity we measured I-V characteristics of DNA-free NaAc and TRIS buffer deposited and dried on the SWNT nanoelectrodes. It was observed that after drying, TRIS buffer exhibits about two orders of magnitude higher current as compared to that for sodium acetate. This observation is in accordance with other recent reports wherein a large current was observed for a similar control experiment with TRIS buffer [4]. Hence it is apparent that, unlike sodium acetate, the intrinsic conductivity of TRIS buffer strongly influences the measured conductivity of DNA and thus can be misled as the intrinsic conductivity of the dsDNA even in the dry state.

Later we investigated the influence of pH of the buffer on DNA conductivity. Figure 2 (c) shows the influence of pH variation on the conductivity of the DNA molecule. After the I-V measurement of DNA molecule at pH 5.8, the substrates were washed with NaAc buffer (10 mM) with pH values from 3.5 to 9.3 using the same experimental process mentioned earlier. It can be observed that as the pH is increased there is gradual increase in current signal. After measuring conductivity at pH 9.3, I-V measurements were repeated by washing the devices with descending values of pH. Current signals were similar to the values obtained for the corresponding pH values during the increase of pH. A similar trend in pH dependence of DNA conductivity was reported by Lee et al [18]. The gradual decrease in current signal with the lowering of the pH can be explained by the thermal destabilization of the dsDNA molecules at extreme pH values (high or low). However this does not explain the increase in the DNA conductivity at higher pH values observed in our measurements. Further studies have to be performed to understand the true mechanism of pH influence on DNA conductivity.

The influence of variation of temperature on the conductivity of the dsDNA molecule attached between the nanoelectrodes was also performed (figure 2 (d)). It was observed that as the temperature was increased from 25°C (in high vacuum) the current signal decreased gradually. We believe that eventual evaporation of water molecules from the hydration shell surrounding the DNA molecule, and subsequent change in DNA conformation became predominant factors in this case. Further increase in temperature above the melting temperature (T_M = 75.6°C) resulted in complete loss of signal possibly due to the thermal denaturation of DNA.

To confirm that the current signal was from the immobilized DNA molecule the following control experiment was performed. The devices were treated with DNase I enzyme (37°C, 30 min). $I-V$ measurements taken after the cleaning step showed no current signal (noise in fA range) indicating that the current signal obtained earlier was indeed from the DNA molecule bridging the SWNT electrodes.

CONCLUSIONS

In conclusion, we have demonstrated that SWNT can be effectively used as nanoelectrodes to trap and electrically characterize single-molecule DNA oligonucelotide. Typically, a current signal in the range of tens of pA (for 1 V bias) was observed for the oligo DNA molecule (80 bp). Influence of various local environmental factors on DNA conductivity was investigated. Results show that TRIS buffer, with its high ionic conductivity, significantly affected the intrinsic conductivity measurement of DNA molecule as compared to sodium acetate which showed negligible effect. A reversible shift in the current signal was observed for pH variation. These findings may shed brighter lights on the ambiguities related to the electrical properties of an oligo-DNA molecule and would facilitate the development of future DNA-based nanoelectronic and sensor devices.

REFERENCES

1. Hall DB, Holmlin RE, Barton JK. Nature **382**,731 (1996).
2. Fink HW, Schonenberger C. Nature **398**, 407 (1999).
3. Yoo KH, Ha DH, Lee JO, Park JW, Kim J, Kim JJ, et al. Phys Rev Letters **87**, 198102 (2001).
4. Kleine-Ostmann T, Jordens C, Baaske K, Weimann T, de Angelis MH, Koch M. Applied Physics Letters **88**, 102102 (2006).
5. Storm AJ, van Noort J, de Vries S, Dekker C. Applied Physics Letters **79**, 3881 (2001).
6. Porath D, Bezryadin A, de Vries S, Dekker C. Nature, **403**, 635 (2000).
7. Giese B, Current Opinion in Chemical Biology **6**, 612 (2002).
8. Giese B, Amaudrut J, Kohler AK, Spormann M, Wessely S. Nature **412**, 318 (2001).
9. Storm AJ, van Noort J, de Vries S, Dekker C. Applied Physics Letters **79**, 3881 (2001).
10. Tuukkanen S, Toppari JJ, Kuzyk A, Hirviniemi L, Hytonen VP, Ihalainen T, et al. Nano Letters, **6**, 1339 (2006).
11. Guo X, Small JP, Klare JE, Wang Y, Purewal MS, Tam IW, et al. Science, 311356 (2006).
12. Roy S, Vedala H, Roy AD, Kim D, Doud M, Mathee K, et al. Nano Letters **8**, 26 (2008).
13. Tuukkanen S, Kuzyk A, Toppari JJ, Hytonen VP, Ihalainen T, Torma P. Applied Physics Letters **87**, 183102 (2005).
14. Kim DH, Huang J, Shin HK, Roy S, Choi W. Nano Lett **6**, 2821(2006).
15. Chen J, Hamon MA, Hu H, Chen Y, Rao AM, Eklund PC, et al. Science **282**, 95 (1998).
16. Chrisey LA, Lee GU, O'Ferrall CE. Nucl Acids Res. 3031 (1996)
17. Kleine-Ostmann T, Jordens C, Baaske K, Weimann T, de Angelis MH, Koch M. Applied Physics Letters **88**, 102102 (2006).
18. Lee JM, Ahn SK, Kim KS, Lee Y, Roh Y. Thin Solid Films **515**(2), (2006).

AUTHOR INDEX

233

SUBJECT INDEX

absorbent, 179
actuator, 221

B, 129

C, 25, 49, 99, 151, 165
catalytic, 25, 37, 157, 215
chemical
   synthesis, 75
   vapor deposition (CVD)
      (chemical reaction), 37,
          129, 157, 173
      (deposition), 3, 11, 151,
          191, 215
coating, 191
core/shell, 55

defects, 17
devices, 221
dopant, 135

electrical properties, 43, 135, 173,
     185, 227
electrodeposition, 55
energy storage, 11

field emission, 129
film, 123
flame synthesis, 67
fluid, 207

liquid crystal, 105

magnetic, 91

nano-indentation, 115
nanoscale, 11, 105, 115, 123, 185,
     221, 227
nanostructure, 3, 17, 31, 37, 43, 49,
     55, 61, 67, 75, 115, 143,
     151, 157, 165, 173, 179,
     207, 215
Ni, 143

optical properties, 43, 91, 105, 185
optoelectronic, 197
oxidation, 99, 135

Pt, 197

Raman spectroscopy, 17, 31

sensor, 227
Si, 123, 143
simulation, 179
stress/strain relationship, 207

thermal conductivity, 165
thin film, 61, 91, 191
transmission electron microscopy
    (TEM), 3, 25, 31, 49, 75

x-ray photoelectron spectroscopy
    (XPS), 99

Zn, 61, 67
zone melting, 197

235

Printed in the United States
By Bookmasters